DIFFERENT SHADES OF AGRARIAN *Crisis* 2021
—Farmer's Distress

DR. HARVINDER KAUR

INDIA • SINGAPORE • MALAYSIA

Notion Press

No.8, 3rd Cross Street,
CIT Colony, Mylapore,
Chennai, Tamil Nadu – 600004

First Published by Notion Press 2021
Copyright © Dr. Harvinder Kaur 2021
All Rights Reserved.

ISBN 978-1-63850-889-2

This book has been published with all efforts taken to make the material error-free after the consent of the author. However, the author and the publisher do not assume and hereby disclaim any liability to any party for any loss, damage, or disruption caused by errors or omissions, whether such errors or omissions result from negligence, accident, or any other cause.

While every effort has been made to avoid any mistake or omission, this publication is being sold on the condition and understanding that neither the author nor the publishers or printers would be liable in any manner to any person by reason of any mistake or omission in this publication or for any action taken or omitted to be taken or advice rendered or accepted on the basis of this work. For any defect in printing or binding the publishers will be liable only to replace the defective copy by another copy of this work then available.

Dedicated to my revered father

Late Ch. Gurnam Singh

Review

This book on agriculture sector in India is a timely publication, which has tried to understand the agrarian crisis in India. This volume examines the transitions in Indian agriculture since 1966s. Comprehensive in analysis, this book has a great relevance, as the longest ever farmer's agitation is going on. It is well written masterpiece, which would be highly beneficial for the readers to grasp the issues, in a comparative manner. The author's attempt to analyse critically the Atamnirbhar Bharat Abhiyan is appreciable and useful to scholars and researchers of economics, political economy and public policy.

Dr. Pardeep Chauhan

Contents

Acknowledgments		9
Foreword		11
Preface		15
1.	Performance of Agriculture Sector: from the Beginning (1966) to Agrarian Crisis.	21
2.	Contribution of Farmers in this Chronic Time of COVID-19	29
3.	The New Policy Paradigm for Agriculture: A Review of Three Farm Acts	33
4.	Agriculture Crisis and Increasing Suicide Rate Among farmers of Haryana and Punjab.	37
5.	Dual Policy of Central Government on Three Acts and Dilemma of Farmers.	43
6.	Contribution of Agriculture towards Self-Reliant India.	49
7.	Crash of the Indian Economy but Agriculture Proved to be the Lifeline of the Economy.	52
8.	Demand Generation by Middle Class and Agriculture sector to Tackle Depressed Economy.	57
9.	MSP is must as a Safety Shield for Farmers.	61
10.	Reverse Migration and Preparation of Source State for this unemployed workforce	66
11.	Can India Become a World Vendor: A COVID Time Opportunism	70

12.	Cost Benefits analysis of Trade with China 'The Global Vendor'	74
13.	Economic Inequality: The Other Side Of Pandemic	78
14.	The Unfolded Story of Health Sector: When Wealth Decides Health.	82
15.	Need of a Paradigm Shift in Economics	87
Conclusion		*93*
Author's Bio		*101*

Acknowledgments

This book would not have been possible without the help of so many people in so many ways. First of all I express my deep reverence, gratitude and indebtedness to my esteemed teacher and mentor and Dr. T.R. Kundu, Professor Emeritus a reputed personality in the field of economics in north India, who inspired me a lot to write this book and gave his valuable time and guidance for completion of this book in the hard time of lockdown and after lockdown.

I also pay my sincere gratitude to our worthy Principal Dr. Rajinder Singh who always motivate and inspire us to do academic works and provides us the required environment and facilities in the institution and it is due it his academic vision I completed this work.

I also pay my sincere gratitude to Dr. Ashutosh Angiras, Associate Professor, Head Department of Sanskrit and who guided me time to time during the drafting of this book and gave his valuable suggestions. I am also grateful to Dr. Balesh Kumar, Librarian of our college for providing valuable resources, whenever I needed and his support in compiling this book.

I am thankful to my loving daughter Sanjivani and my husband Rajinder Singh for their cooperation and care during the time of writing this book.

Foreword

Prof. (Dr) T R Kundu
Prof. Emeritus,
Department of Economics,
Kurkshetra niversity,
Kurukshetra.
Haryana

Indian agriculture today is at critical juncture, with farmers the Central govt. and farmers taking diametrically opposite stands. The govt. has come out with trio farm laws claiming them to be panacea for the country's ills, while farmers dubbing the same as the proverbial remedy worse than disease.

The present agrarian crisis has not appeared suddenly on the scene rather it's genesis can traced to be the mid 1980's when the green revolution technology began to exhibit diminishing returns. The water and chemical intensive technology had taken heavy tolls of natural resources in terms of depletion of ground water level and damage to the soil health. It was high time for adopting crop diversification and eco-friendly technology. However the policy makers didn't pay attention to it. As a matter of fact the government became indifferent to the agriculture sector, after having achieved self sufficiency in food grains in the wake of green revolution. Even the major policy reform of 1991 did not cover agriculture sector. No wonder that the conditions in agriculture have worsened during the post reform period, with both the public and private investment showing downwards trend and agriculture growth sliding to new low levels. There is a need to understand the agrarian crisis in full dimensions and make appropriate policy intervention, opening up of new vistas of agriculture growth and prosperity for farmers.

The present volume by Dr. Harvinder is a welcome contribution in this direction. The author forays into aspects of the crisis and suggests measure to deal with it. She forcefully advocates a comprehensive regimes covering all the major crops. It would encourage the must needed Crop-Diversification, avoid wheat and rice and reduced imports of pulses and edible oils. The book is slim in size but rich in content. It's simple language and engaging style are some other distribution features. In all, it is a worth reading book for the academician, policy makers, social activists and for the public at large.

Foreword

Ashutosh Angiras (Dr., Capt)
Hony Director,
SDHDR & T Centre
Sanatan Dharma College,
Ambala Cantt.

Economics in itself is ever changing and expanding activity with which all economists have to make special efforts to keep up pace with its dynamic nature. One cannot assume that by knowing fundamental laws of economics will be sufficient to define economic behaviour and its pattern. World over economists are constantly striving for reframe economic behaviour with all its dimensions – be it wellbeing economics or green economics or sustainable economics etc. Issue is how deal with agrarian economics and market economy, how to synchronize them in such a way that both not only mutually enrich each other but also serve human wellbeing. It will be good idea to talk about sensible economics now when world over market economy is dictating its terms and along with it conditioning human behaviour including thought process. Economics needs a visionary mind which is capable of dealing human wellbeing intellectually and intelligently. It must pursue sensible economics where society and individual is free from compulsive choices and becomes capable enough to make conscious choices by a single line statement – this much is required and no more is needed. But as we all know this is not the case right now.

Foreword

The author of this book Dr Harvinder Kaur, Assistant Professor in Economics is an emerging critical thinker of Economics and its allied fields who can see through the economic policies, their future impacts on society particularly agrarian economy. Her academic flavour can be felt in her multiple articles published by reputed newspapers & journals. Her sole concern is human wellbeing through restructuring agrarian economics. Her analysis and insights in the agrarian primary, moving and dangerous contradictions are worth appreciable. Her critical review of the present day three Farm Acts and farmers agitation is appreciable. As an emerging economic thinker I welcome her line of argumentative thinking.

Her articles on various aspects of economic concerns, their analysis and suggested way outs are an attempt to bring a paradigm shift in upcoming economic form and essence.

This book is compilation of different articles written by the author in different newspapers during the period April 2020 to February 2021 and is divided in two sections, first section deals with issues of Indian agrarian economy. In this section the importance of agriculture sector has been presented and problems of agrarian economy has been raised in a factual manner. Second section is deals with the area of opportunism to transform Indian economy into a global factory and along with it cost benefit analysis of trade with China has been discussed. Increasing economic inequalities among the population and health related issues have been raised which were at the worst stage during pandemic times. Due care has been taken by the author to frame the issues or problems in the right and just manner and then sincerely tried to decode them as a sensitive economist.

I wish her all the best for her future academic endeavours.

Preface

The story of north India's growth in agriculture sector begins after 1966 when a strategic decision was taken and high yielding varieties of seeds were used with chemical fertilizers in production of wheat and rice. The credit goes to Norman Burlog a American scientist and after this green revolution production of wheat and rice increased multiple times. After independence, there was a need for agricultural practices to be redefined so as to feed the ever-growing population of India and reduce dependence on import of wheat. For the same, during the time of plan holiday in 1966, in Five Year Plan to improve farm productivity, To achieve the target of food security technology was introduced in agriculture with HYVSs and an incentive was given to farmers to produce wheat and rice in the form of MSP and we achieved best results as cereal production was 131 million tons in 1978-79. This established India as one of the world's biggest agricultural producers. Yield per unit of farmland improved by more than 30% between1947 (when India gained political independence) and 1979. In1985 diminishing returns started in this sector and after this, this sector was totally ignored and public and private investment declined., public sector investment, was 33 per cent of the total investment in agriculture sector in 1985, was the main reason of the Revolution, but now (2019) it has reduced to only 17 per cent and after 1991 and because of the pressure of WTO this sector was totally ignored and presently govt. want to remove APMC and unbundle FCI and want to enter private corporates in marketing of agricultural production.

This book is compiled of different articles written by author in different newspapers during the period April 2020 to February

Preface

2021 and outcome of various changes in economy during the prolonged lockdown in Indian economy due to COVID-19,as farmer's protest over three bills, migration and reverse migration of labour, conflict on MSP between farmers and government, increasing suicide rate among farmers, farmers indebtedness, increasing income inequalities and many more incidences, which took place during pandemic. Although lockdown has positive impact on environment but because of this, economic environment got ruined of already battered economy.

Indian economy is a developing economy having 14% of total population of the world with 4% area of total land of the world with a growth rate of 6.8% in 2019 (Pre. COVID). Agriculture plays a major role in the economy as about 57% population directly or indirectly depends on this sector for their livelihood and this sector contribute about 17% in GDP. India's geographical conditions are growth stimulating as this is the only country after America having maximum arable land.

This book is divided in two sections. The first section deals with the emerging problems in agriculture sector of the economy. There are many factors which are responsible for this low growth rate as farm sector depends on monsoons, inadequate irrigation facilities, lesser use of machines and inadequate policies of each government for farmers. All these factors have made agriculture completely a profession of loss. In the past 20 years, about 3.2 lakh farmers have committed suicides in India. In absolute numbers, Farmer's suicide in Haryana and Punjab were less than in Maharashtra which is on number one. Punjab is the latest to emerge as a farming graveyard, in 2015, as many as 449 farmers had committed suicide. Punjab and Haryana, the food bowls, has now the second highest rates of farmers suicide in the country. This is baseless that farmers suicides are due to lack of irrigation or low productivity but hidden truth is that, the denial of appropriate price of crop is the main reason for the increasing suicide rate in big agrarian states which have more than 80 percent land has irrigation facilities.

There are about 15 crore farmer families in India and out of these about 52% (NABARD REPORT) farmers are under debt. According to a study of WTO (2018-19) every farmer receive a subsidy of Rs.20,000 from all resources. Centre and state government together gives a subsidy of 3.25 lakh crore in the form of seeds, fertilizers, MSP, electricity etc. but on the same time industries has been given a subsidy of 10 lakh crore and in America 45.22 lakh subsidy is given to each and every farmer.

On an average each farmer has a debt of one lakh rupees and this debt is increasing in Geometric Progression and income of a farmer is frozen for the last 40 years. About 28 farmers commit suicide every day in India. On an average income of a famer family according to the report of NABARD is 8,931 rupees per month.

The unpredictability of agricultural yields and products often results in windfall losses making the lives of farmers debt-ridden and pathetic. The economic decisions involved in agriculture are massive and need to be recognized as such. The demand of Minimum Support Price (MSP) is directly linked with food security in India. Hence, farmers have a rationale and urgency in seeking govt attention towards their economic well-being as made or marred by new agricultural legislations of the central govt. For instance, students of economics are familiar with the COBWEB theorem. A farmer have to take decision to produce particular crop minimum six months before marketing so then a assurance of minimum support price is must in agriculture. Every industry can increase or decrease its production according to the demand in the market but a farmer can not as he has sown the crop six months before, it reach to market and standard law of demand and supply of Dr. Marshall does not apply in food production.

In this section I have try to justify the demand of agitating farmers for a legal provision of MSP. As for as Minimum Support Price is concerned, MSP of crops is fixed very low deliberately to curb inflation, so that industries can get cheap raw material and to attract multi national companies in India and to reduce fiscal deficit of the government. Control inflation is a praise worthy step but this is unfair to put all burden of growth on the shoulders of farmers, who are enough innocent. Why the farmers should be penalised for growth by curbing inflation rate of food basket ? According to Swaminathan report, MSP should be C2+50% but farmers are deprived from their rights.

Only 38% cultivable land have irrigation facilities and remaining is rainfed, so increase the possibilities of windfall gains and losses in agriculture and leave the farmer on the mercy of market forces of demand and supply is not a praiseworthy action.

Income inequalities which were earlier high and increased in pandemic times.. There are 119 Billionaires in India. There number increased from 9 in 2000 to 101 in 2017. Between 2012 and 2018, India is estimated to produce 70 new millionaires every year. The question arises, is Make In India Abhiyaan making millionaires? In India, the wealth gap has been rising sharply during the pandemic as previous level of income inequality was already high in the economy. 23 crore people in India sleep hungry every day. Our Global Hunger Index slips down from 102 rank to 94th rank out of 107 Countries. India's rank takes it below Bangladesh and Pakistan.

2nd section is About the area of opportunism before India as China has proved itself villain by spreading fake news and misguiding the world about Corona virus. Can India fulfil it's dream to be a world vendor and can take benefit of being the home of maximum young population of the world ?. Can it attract more FDI and can it take the place of China ?. Both economies started their journey of development in 1950. But now India is a 2.9trillion economy and China is 14 trillion economy. China do skill based work where per worker value added is multiple times greater in comparison of India. China's value added per worker in manufacturing is the biggest evidence of it's growing technological powers. So this is not a easy play for India to become global factory.

As for as trade with China is concerned,our pharmaceautical industry depends on China for supply of ingreniants for the production of final goods. India depends on China for 80% of its supply of active pharmaceutical ingredients (API). Overall, China makes half of the planet's API, according to Britain's Medicine and Healthcare Products Regulatory Agency and pharmaceutical analysts. We have to do big efforts to search the economic substitutes of China's imports. Than how the Atamnirbharta mission of India can be fulfilled in these dwindling situtions.

The government aims to raise the public health expenditure to 2.5 per cent of GDP by 2025 in a time-bound manner but this expenditure is not sufficient as it should be minimum 4 to 5%

of GDP. Presently India is spending only 1% of it's GDP on health and we are far behind of many developing countries. In budget 2020-21 government has given a proposal to increase the budget of health 137%. Let's hope that this all should not be only on papers and this should be a step towards the mission of healthy India.

In last chapter, suggestions for a paradigm shift in Indian economic policies are suggested, so that these multiple problems can be solved which arise after lockdown and pandemic. To return to the previous path in a competitive world we cannot ignore interdependence of international trade as manufacturing sector of India depends on imports of comparatively cheap economic substitutes for the production of final goods. We should never forget that to achieve the target we have to adopt a flexible policy of liberal trade. If growth is necessary then income equality is must for a democratic country like India as someone said that poverty anywhere is a threat to prosperity everywhere.

Instead of ANBA (Atamnirbhar Bharat Abhiyaan) and self - reliant India, our strategy should be of balanced growth, having equal importance of every sector.

Performance of Agriculture Sector: from the Beginning (1966) to Agrarian Crisis.

The story of north India's growth in agriculture sector begins after 1966 when a strategic decision was taken and high yielding varieties of seeds were used with chemical fertilizers in production of wheat and rice. The credit goes to Norman Burlog a American scientist and after this green revolution production of wheat and rice increased multiple times. After independence, there was a need for agricultural practices to be redefined so as to feed the ever-growing population of India and reduce dependence on import of wheat. For the same, during the time of plan holiday in 1966, in Five Year Plan to improve farm productivity, seven districts chosen from seven states during 1966-67, the Government started an ambitious development plan, in the form of Green Revolution in India. To achieve the target of food security technology was introduced in agriculture with HYVS and an incentive was given to farmers to produce wheat and rice in the form of MSP and we achieved best results as cereal production was 131 million tons in 1978-79. This established India as one of the world's biggest agricultural producers. Yield per unit of farmland improved by more than 30% between 1947 (when India gained political independence) and 1979. Green Revolution was a major event in the 1966-69 it boosted agricultural productivity, almost doubling it with the help of technology and high yielding variety of seeds along with mechanization and industrialization of traditional agricultural activities. Country's foodgrain output is set to increase from 285.2 million tonnes in 2019 to 291.95 million tonnes in 2020.

The rice production is likely to increase at all times high of 117.47 million tonnes.

The Green Revolution in India began in 1965 under the leadership of Sh. Lal Bahadur Shastri, which led to an increase in food grain production, particularly in Punjab, Haryana, and Uttar Pradesh. There was a need of green revolution to feed the rapidly increasing population and to increase agricultural productivity.

Before this we used to import wheat from America under P.L 480 scheme on subsidized rates,but in return of this America used to impose upon us political and market conditions. In initial stage before and after independence rice and wheat were not included in food items of north Indians. North Indians were called brown eaters (wheat)and south Indians were called 'white eaters' (rice). Rice and wheat were included in luxurious food and were out of reach of common people. But now the Punjab and Haryana are called bowl of rice. In 1966 the area under rice and wheat production was purposely selected under a strategy. Karnal belt of Haryana (including Kurukshetra) was selected for a special variety of paddy called Basmati (name comes from "mitti ki baas") and first seed of Basmati was sown here. But now this bowl of rice belt is under water crisis and some zones has been declared dark zones where ground water level has gone deep.

Numerous studies shows that after mid 1980s diminishing returns starts in agriculture in green revolution area because of many factors. It is no longer sustainable, environmentally as well as economically as water level is coming down and production cost has also increased. Public as well as private investment has decreased in agriculture. During the Green Revolution, public sector investment, was 33 per cent of the investment in agriculture sector in 1985, was the main reason of the Revolution, but now (2019) it has reduced to only 17 per cent. In absolute terms, the real agricultural investment during the year 2016-17 was 2.6 lakh crores accounting to about 2.2 per cent of GDP. The 83 per cent of the investment by the private sector, primarily by farmers and the contribution of the corporate sector is only 2 per cent. As the investment capacity of the farmers is very low as 80 % of them are small and marginal farmers, being so public sector investment should be substantially enhanced. The level and scale of public investment has direct impact on overall GDP growth. The public sector investment should steadily be enhanced to accelerate the overall GDP growth rate, which will directly help in increasing farmers' income. The corporate sector should be boosted to strengthen R&D, inputs supplies like quality seeds, agrochemicals, machinery, processing, and value addition.

Lack of public as well as private investment was the main reason of diminishing returns with many other reasons as soil degradation and depleting water level. After the liberalization 1991 the agriculture sector was totally ignored and tenancy trend started in agriculture and private investment by the farmer declined and now a new system started in agriculture reverse tenancy.45% farmer want to leave this occupation as according to them this profession has now become a profession of losses. The time period after 1985 was the main time of crop diversification but at that time government neglected this sector.

In Haryana and Punjab paddy irrigation in initial stage is done with water pumps and that is issue of concern from 1985 and has increased ground water table problem. That's why farmers are instructed to plant paddy after June 15. So alternative crop pattern is need of hour in place of paddy and wheat and government can make it feasible by announcing MSP for all crops. This year Haryana government has launched "mera paani meri virasat" scheme to promote crop diversification. Farmers who will diversify crop on 50 percent land will be given an initiative of Rs.7000 per acre. In this way we can encourage sustainable agriculture and can maintain sinking water table for future generations. If we try to find the reason behind more paddy production is the assured market of paddy under MSP and as a result non-rice product areas have taken paddy production on a large scale where maize used to be a major crop before 1970. Maize requires about 4 to 5 water cycles while rice requires 40 to 50 irrigation cycles, hence affects groundwater level.

There is need of crop diversification and incentive in the form of MSP should be given to farmers so that they prefer crop diversification. As government of Haryana promise to pay rs. 7000 per acre for crop diversification in the form of crop compensation for replacement of paddy. This crop compensation (fasalantar) is given before sowing of crop while Bhavantar Bharpayee scheme work after production at the time of marketing when market price of crop is less than support price. These both schemes should work parallelly with assured MSP to induce farmers for crop diversification as these schemes are the immediate and short run solution to per mote diversified crop pattern in replacement of paddy and wheat. A commission should be made to investigate the problems of farmers and to make them assure about the long run benefit of crop diversification and aware them, so that they can take benefit of these schemes.

India currently is the largest exporter of rice, contribute nearly 25 percent of total agriculture exports. But because of this depleting water level crop diversification has become need of hour and if government implement it technically and strategically, it can bring a new green revolution. Presently India's government is spending only 0.3 percent of GDP on research for agriculture which is less than many developing economies of the world, so there is need to spend a large proportion of budget on agriculture research and crop diversification. As government of Haryana is taking initiative to increase the area under sugar cane production for the purpose of crop diversification but government have to build infrastructure for this purpose, existing sugar mills capacity will have to be expand with establishment of new sugar mills. Farmers should be convinced and initiative should be given to them for crop diversification as chief of Haryana kisan union Gurnam Singh Chaduni said that government has also promised to pay Rs. 2000 per acre for crop diversification but amount was not paid. MSP is another tool with Bhavantar Bharpayee yojana and Fasalantar Scheme to promote crop diversification. Maximum crops should come under MSP and it should be declared timely so that farmers can take decisions regarding sowing and marketing of crops. MSP should fixed at least 50 percent higher than the weighted average cost of production and in case of diversified crop sit should be atleast 75 percent. In paddy production government can give subsidies to farmers for furrow irrigation system and drip irrigation. Furrow irrigation method is different from flooding irrigation method and is used in many developed countries for paddy irrigation. Digital technology can also be used to aware the farmers to tell them appropriate planting time of paddy and about monsoon arrival so that farmers can prepone or postpone water cycle time and can save ground water.

Water harvesting can also be helpful in irrigation.

Large budget is required for rain water harvesting, this can be useful in irrigation of paddy in month of may and june. Rain water harvesting need to be implemented in rural areas to recharge ground water level.

Government have to create a assured market and assured price with a research based technology. It is need of hour for a dynamic economy as RBI governor said in a press conference on 22 may that agriculture sector is the sole hope in depressed and battered Indian economy.

Now a days a new crisis has emerged in agriculture sector because of three new bill for farmer and farming. Lakhs of farmers are agitating on Delhi borders. We will discuss about them in detail in next chapters.

According to United Nations World Food Program (WFP), the temporary close down of economic activities has doubled the number of people facing acute food insecurity across the world to 265 million. Mr Arif Hussain, Chief economist at UNWFP, stressed in his speech at Geneva that we all need to come together to deal with these uncertain times because if we don't, then the cost will be high as 'Many will lose their lives and Many more will lose livelihood'. IMF has predicted that the global economy will experience the worst global downturn since the Great Depression, with the global growth in 2020 expected to fall to -3 percent.

As far as India is concerned, this pandemic has broken the backbone of already battered economy. According to IMF, Indian economy will grow at the rate of 1.9%. The Indian Economy is expected to lose over Rs 32,000 crore everyday during the first 21 days of nationwide lockdown. Unemployment rose from 6.7 percent in March 2020 to 26 percent as on 19 April 2020. Nearly 14 crore people lost employment within a month and economy crash upto 24%, only the agriculture sector show positive recovery with 3.4% growth. According to ILO, India will add 300 million persons to the number of poor. About 53 percent of businesses in country will be significantly affected. There will be widespread joblessness. Unemployment will increase. The last hope in these uncertain times for India is it's agriculture sector. FCI has assured that India has abundant food grains and there will not be any shortage of food grains in our country. Even in these unprecedented times, the Indian farmer continues to serve their countrymen. Country's foodgrain output is set to increase from 285.2 million tonnes in 2019 to 291.95 million tonnes in 2020. The rice production is likely to increase at all times high of 117.47 million tonnes. It is these abundant foodgrains that will keep us going in the war against this pandemic.

But this also presents itself with a situation akin to the economic problem of market dynamics. These abundant foodgrains may deteriorate the prices of the food grains and as result farmers and agriculture sector, which is the lifeline of the economy, may collapse. Due to the decrease in prices, farmers may not be able to recover their cost of production. Moreover, the migrant labour has also fled to their home states and this has added to the farmers' woes in the peak season. This is tough time for Indian farmers because if this continues for a long time, the day is not far when they will also join the army of labourers and the agriculture will become the occupation of loss. It is now more urgent than ever to promote agriculture by making it a Profession rather than an Occupation. Agriculture sector should be promoted as an Industry whereby providing with all the necessary help. Farmers should be encouraged to produce Commercial crops as well as foodgrains so that they can earn profits, and render to the needs of growing population and Industries as well. There is a need for the new agricultural policy that can bring the revolutionary changes in the sector. Government should set up a commission to analyse the conditions of farmers and to settle disputes of farmers and commission agents. The migrant labourers who went back to their home states should also be gainfully employed in small and medium scale industries so that they can become wage earners and not the disguised labourers of agriculture.

The agriculture sector in India have been ignored for long time but this needs to be changed because if nation's economy were a human body, then it's heart would be the agriculture sector.

Contribution of Farmers in this Chronic Time of COVID-19

**Nahi hua hai abhi savera, purav kilaali pehchaan!
Chidiyon ke Jagne se pahle, khaat chod uth gaya kisan!!!!**

As far as I remember that I read this poem written by Ramdhari Singh 'dinkar' when I was in 4th or 5th class in my village high school. Whenever I recited these lines in my younger age, image of my father emerged in my mind. My father was a farmer in a small village in district Kurukshetra (Haryana). I always saw that he used to wake up very early in the morning with the chirping of sparrows, birds and used to go his farm for his usual farming activities. Once he shared with us sense & understanding regarding how to estimate time of early morning with the movement of stars in the sky when clocks were not available to common people. That was really a magic trick and always mesmerized me. Slowly I learned the technique and used that method for the purpose of experiments fun sake. Today too I enjoy to play this art with my children.

I hardly remember that my father had any breakfast ever. His breakfast, lunch used to in the fields only, around 11:00 a.m. with his labourers. Though this 11.00 a.m. was not a fixed meal time, he used to eat it during the available free time amid the field work. He was an enlightened & progressive farmer who knew the importance of higher education for a girl child.

I am sharing this story of my childhood as I wish to make everyone aware about the life of a farmer which has not changed till date in India as well as in Haryana state.

In today's scenario when this pandemic COVID-19 is spreading in whole world including our nation also, a farmer still goes to his fields/ farms as usual. At the time when the whole world is affected by this epidemic and we all are locked in our houses for the safety of our lives with enough storage of food grains, cereals etc, and a farmer goes to his daily field work to produce food grains for all of us without any hitch. Isn't it strange to realize that our priorities are only to secure food grains for ourselves and not for others? A farmer is doing for others, for the whole society.

In 1960 late prime minister Sh. Lal Bahadur Shastri gave the slogan "Jai Jawan, Jai Kisan", which still proves to be very accurate and meaningful today too.

April is a time of crop cutting of wheat in north India. The crop will be brought to grain markets by farmers. Farmers are not going to hoard it for the purpose of black-marketing to earn big profits as all other dealers generally do because farmers are loyal and innocent by nature. They always say with proud that they are the son of soil and respect earth as their mother.

Growth rate of agricultural sector in India is very slow in comparison to all other sectors. There are many factors which are responsible for this low growth rate as farm sector depends on monsoons, inadequate irrigation facilities, lesser use of machines and inadequate policies of each government for farmers. All these factors have made agriculture completely a profession of loss. But still the farmer is engaged in agriculture activities for the purpose to serve his nation, to feed 130 crore population.

As far as the state of Haryana is concerned, the situation is better than all other states. As per economic survey of 2019, Haryana constitutes 1.5 percent of total area of India's total land yet contributes 15% in total agriculture production. Its productivity has increased 7times since 1966 when this state came into existence. Agriculture sector contributes 16% to GDP of Haryana, having less contribution in comparison of other major sector industry and services.

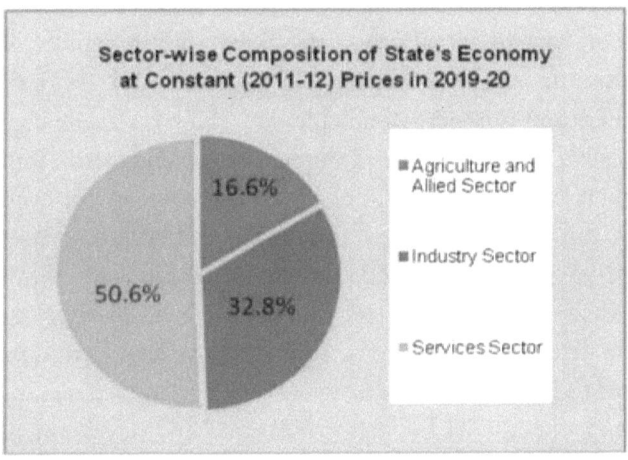

In agriculture sector one can see a paradox which we call "poverty in abundance" in economic terms. And this is true in the state of India as well as in Haryana agriculture sector. This is the fate of farmer. After producing in abundance he still remains poor because market conditions are not in his favour. But this time is not suitable to talk about profits and losses. Let's fight with this pandemic with courage and with sufficient food grains to eat for all our citizens.

We would have lost the race against COVID-19 if we have not sufficient foodgrains to feed our population and still the government is ignoring the role of farmers and imposed three ordinance on farmers of trade and commerce and contract farming to protect the interest of private corporates. Centre government is designing policies according to the global design of WTO and forget that in our economy there are more than 60 crore farmers. Government is going to abolish MSP, FCI and FSA according to the suggestions of Shantakumar committee and agriculture sector will ruin and this pandemic time was not suitable to bring these ordinance, as government has brought these ordinance from the back door when the session of parliament was not going. Abolishment of MSP (minimum support price) will increase distress among farmers and can increase suicide rate which is already high in states like Maharashtra and Punjab. According to an estimate farmers in 17 states farmers are surviving on an average only earning Rs.1700 per month. For the last six years-real income of the farm sector is almost frozen with a minor increase of 0.44 percent but the farmer is still struggling in these deteriorate conditions. So the government should think about the revolutionary changes in the agriculture sector and should increase it's budget on Research and development of agriculture sector from a minor expenditure of 0.3 percent of GDP to about 10 percent of GDP. To double the income of the farm sector (as government has already promised) upto 2022 the growth of agriculture sector should be 10 percent and in upcoming two years the agriculture will prove the lifeline of the economy as all other sectors are having declining rate of growth because of pandemic. So to save the lifeline of the economy the economic policy design should be accordingly and should promote agriculture sector as an industry.

The New Policy Paradigm for Agriculture: A Review of Three Farm Acts

Government has already accepted that it is not possible to double the income of farmers upto 2022 and ultimately it has removed it's mask of being well wisher of farmers and dropped the idea of double income in doldrums. To double the income of farmers, farmer's income should grow at least 10 percent per annum, but in these dwindling situations it has become difficult for the farmers to survive and to continue their occupation of farming.

In 1980 growth rate of agriculture was 5.7 percent greater than growth rate of economy which was 5.5 percent because of huge expenditure on irrigation. In ninth plan it was 2 percent and in 10th was 1.8 percent.

The annual growth rate in real terms in agriculture and allied activities was 2.88 percent from 2014-15 to 2018-19, almost constant for the last six years and this all because of reduction in public investment in this sector, as maximum money for reinvestment comes from taxes of middle class. Policy design of government is to reduce the number of farmers from 57 percent to 30 percent.

Swaminathan report says that the growth in agriculture should be measured not only with increase in productivity but also with the 'net take home income 'of the farmer and the income of the farmer should be comparable to those with civil servants but for the last 6 years real income of farmers has been frozen with a minor increase of 0.44 percent. Agriculture sector is the backbone of the economy and there are two main reasons behind this, first 60 crore farmers are not only producers of food grains and raw material but, also a major component of demand chain. But the policy of the government has made them the supplier of food and cheap raw materials, as well as suppliers of cheap labour to industrial sector because of the global design of WTO and it's pressure on central government.

Mahatma Gandhi once said "To those who are hungry, God is bread." A majority of the hungry, lives in rural India and they depends on agriculture for their livelihood but the centre government wants that the farmers should migrate to urban areas as supplier of cheap labour.

According to the recommendations of Shanta kumar committee, which presented it's report in 2015, the MSP, Food security Act and Food corporation of India should be abolished and government should allowed private corporates to collect, store and to distribute foodgrains in all India. This committee gives the logic that initially Punjab, Haryana and Utter Pradesh produced wheat but now wheat is produced in whole India, so it is not possible to give MSP to the farmers of whole India for wheat and other foodgrains. This committee also recommended to abolish Agriculture Produce and Market Committee.

Can the government explain that during lockdown farmers throw their vegetables, fruits and milk on roads and no private corporates came to help them rather it was the cooperative milk diaries that bought surplus milk to convert it into milk powder and cheese, as Amul and Verka in Punjab and Vita in Haryana. Three ordinance of government recently imposed on farmers are according to the suggestions of Shanta Kumar committee, First bill is about the amended Essential commodities Act to remove the existing restrictions on some crops as potatoes, onion, pulses and some other crops. Private corporates can store these crops for the purpose of processing, this ordinance is not for the benefits of farmers but rather for the benefit of private players because if government want to help farmers than it can instruct private firm to make farmers stakeholders in processing, and when the value of agricultural products will increase after value addition then the farmers will be benefited and their income can increase but the government has no such intention. If government make farmers stakeholders in value addition then agriculture will become a profession, instead of occupation and investment will be done like industries.

Second bill is of trade and commerce, as according to this ordinance farmers can sale their crop anywhere in India but the government forget the ground reality that for small and marginal farmers, it is not feasible for them to take their produce as they have not any resources of transportation moreover they have taken debt from money lenders and they are bound to sale their crop only to them, so this ordinance too have not benefited farmers rather this ordinance is also for the benefits of corporates.

Third bill is more dangerous as according to this APMC and FCI will be abolished and private firms will contract with farmers to purchase their crop on a contractual price but under the cover of contract farming government is playing game with farmers as there is fear that private firms will blackmail and bargain with farmers by making cartels. There is every possibility that the private firms can force farmers to sell their crops on low prices by targeting them that the crop is not according to the standard of deal price. MSP will also be abolished as it has not make clear by the government that the private firm will purchase the crop on MSP, moreover the farmer cannot move to court if he will face any problem, he can make complaint only to SDM. So there is every chance that farmers will be exploited by the private firms.

Government should try to do evolutionary and revolutionary changes in battered agriculture sector. As for as Agriculture Produce and Market Committee is concerned, centre government want to abolish it and want to unbundle FCI and allow private players to procure, store and distribute food grains. In this way government will reduce it's storing and tiring cost.

More than 50 percent population depends on agriculture sector and agriculture contributes about 18 percent in GDP but only 0.3 percent of GDP is spend on agriculture research and development, which is too low in comparison of developed countries. India is the second largest producer of wheat and rice, the world's major food staples. India is currently the world's second largest producer of several dry fruits, agriculture based raw material, roots and tuber crops, pulses and farmed fish, eggs, coconuts, sugarcane and numerous vegetables. So a huge public investment in agriculture machinery and on research and development is need of hour to make agriculture sector an industry. Rs.75000 crore allocated to farmers in the budget of 2019-20 should be increased in this COVID time, as direct cash transfers to farmers. Number of APMC should be increased, so that more farmers can sell their crops in regulated markets and can be saved from the clutches of money lenders as only 6 percent farmers sell their crops through APMC on MSP according to Shanta Kumar committee report and any firm or person purchase the crop from the farmer below MSP then it should be a punishable offence.

Agriculture Crisis and Increasing Suicide Rate Among farmers of Haryana and Punjab.

Whenever government pretend that it want to solve the problems of farmers, a story comes in my mind that once an old lady was searching her needle under a street light and a passer-by also started searching with her, for the purpose to help her and when both doesn't find it, the person ask the old lady, "are you sure that you lost the needle here. The old lady very innocently answered that I lost it in my hut but there was too dark so I thought let's I search it in light.

The same game is played by the government for agriculture sector and farmers, as the government very well knows the root cause of pathetic state of farmers but pretend to be sympathetic and well wisher of farmers and try to give lollipops to farmers. Let's peep inside the agriculture sector to find the root cause of the problem.

Agriculture scientists and economists have always suggesting to increase crop productivity to enhance the income of farmers and blaming low productivity for the distress of farmers. We are listening from the very starting that in the era of globalization, one can only survive if one becomes globally competitive and those who remain behind are left with no option but have to commit suicide, and same is with farmers.

In the past 20 years, about 3.2 lakh farmers have committed suicides in India. In absolute numbers, farmers suicides in Haryana and Punjab were less than in Maharashtra which is on number one. Punjab is the latest to emerge as a farming graveyard, in 2015, as many as 449 farmers had committed suicides. Punjab and Haryana, the food bowls, has now the second highest rates of farmers suicide in the country.

National Crime Record Bureau (NCRB) has presented the data regarding the suicide of farmers in India. According to this data in 2015 12,602 farmers committed suicide and in 2016 11,370 farmers committed suicide. The reason of lack of irrigation facilities has been forfeited as in Punjab 94 percent and in Haryana 82 percent land have irrigation facilities and still in Haryana in 2016, 250 suicidal cases has been reported in comparison of 162 in 2015,it means suicide rate has increased 54 percent in Haryana. In Punjab, the rise was 118 percent from 2015 to 2016.

FARMERS' SUICIDES
(includes those by farm labourers)

	2015	2016*	% Chg
Punjab	124	271	**118.0**
Haryana	162	250	**54.32**
Karnataka	1,569	2,079	**32.50**
Gujarat	301	408	**35.5**
Madhya Pradesh	1,290	1,321	**2.4**
Telangana	1,400	645	**-54.0**
Maharashtra	4,291	3,661	**-15.0**
Andhra Pradesh	916	804	**-12.2**
Chhattisgarh	954	682	**-28.5**
Total	**12,602**	**11,370**	**-9.8**

Note: Total might not match as all states have not been included Source: Parliament questions

Human activists have questioned the data given to parliament by NCRB as ' under reporting perhapsNCRB has played with data and for the last four years it has not collected any data about farmers suicides.

This is baseless that farmers suicides are due to lack of irrigation or low productivity but hidden truth is that, the denial of appropriate price of crop is the main reason for the increasing suicide rate in big agrarian states which have more than 80 percent land has irrigation facilities, as told by agriculture expert Devender Sharma. He said that Punjab numbers were suspect as data from house to house surveys conducted some years

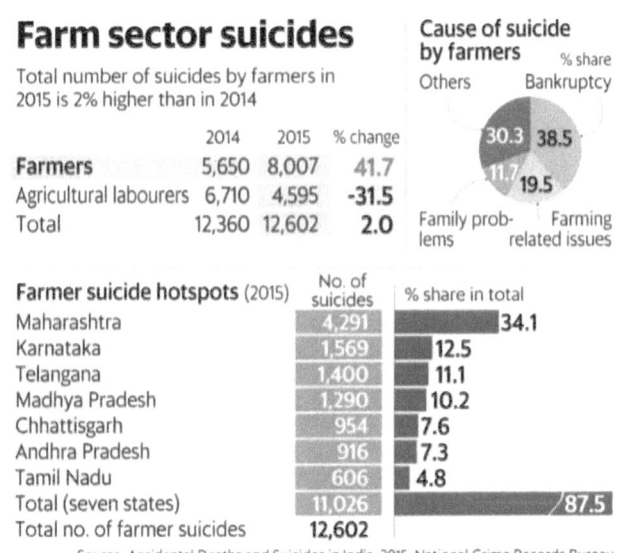

Source: Accidental Deaths and Suicides in India, 2015, National Crime Records Bureau

earlier by leading universities of the state, showed that 16,600 farmers ending their lives between 2000 and 2016 and about or 1000 farmers suicides per year but NCRB figures says it is less than 271 per year. More than 900 farmers have taken their lives during the past two years in Punjab. In January 2019 alone, 32 farmers suicides were reported in Punjab. Crop failure and under price that doesn't even cover the cost of production are the main reasons, so the farmers are unable to repay their loans and goes in distress or commit suicide.

No doubt irrigation facilities and technology advancement increase productivity but there are some other important factors are also here and if we make agrarian crisis only the sole reason for increasing distress and suicide rate of farmers, this will be the very easy solution of the problem. It can be partially applied in Maharashtra as there is only 18 percent land have irrigation facilities but we cannot apply this logic in Haryana and Punjab. Productivity of each and every crop in Punjab and Haryana is not less than the average productivity in developed countries. Punjab (7,633 kg per hectare) leads the productivity chart leaving behind US (7,238 kg), UK (7,460 kg) and 5,920 kg in Japan. If raising productivity is the major reason, we doesn't find why Punjab farmers should be committing suicide. But the fact is that, that government and economists don't want to acknowledge that the low prices of crops (which is deliberately fixed) and market imperfections are the major reasons behind the farmers suicides.

MSP of crops is fixed very low deliberately to curb inflation, so that industries can get cheap raw material and to attract multinational companies in India and to reduce fiscal deficit of the government. To control inflation is a praise worthy step but this is unfair to put all burden of growth on the shoulders of farmers, who are enough innocent, why they should be penalised for growth by curbing inflation rate of food basket. According to Swaminathan report, MSP should be C2+50% but farmers are deprived from their rights. C2 include full cost as rent of land, tractors and imputed value of the labour of family members but government play game here too as it includes only purchasing cost to decide MSP, moreover because of lack of regulated markets only 6% farmers can sale their crops on MSP remaining 94 percent still sold it in unregulated markets because of market imperfections or because of debt ridden situations of farmers. There are less than 7000 regulated APMC (agriculture produce market committees), and India need vast network of about 42000 markets if mandis are provided in 5Km radius. Government is going to bring a new ordinance of contract farming and it will be a big danger as it will deprived farmers from MSP and farmers will be in clutches of private purchaser, and a web is going to be woven for farmers in the form of contract farming. Government will save himself from procurement, storing and trading cost but agriculture sector will be destroyed.

Let's compare the income of farmers of India with other section of the economy.

The procurement price of wheat has increased 19 times over a period of 45 years and if we compare it with the income of various other sections of the economy. The increase in the salary of government employees, college/ university teachers, school teachers is 120 to 150 times, 150 to 170 times and 280 to 320 times respectively and if we follow the minimum yardstick of 100 times increase in procurement price for the same time period, wheat price should be Rs.7600 per quintile. This is the legitimate price of wheat, which has been denied to farmers. No doubt it will increase inflation rate of food basket but a farmer should not be penalised for producing food. And the strange fact is that the farmers of UK and US are not rich because of more productivity but because they are highly subsidised to produce food grains of given standards and direct cash transfers are also given to them by their governments. When the farmer of India sow seeds of crops he knows that he will cultivate losses but he is helpless. Government can pay them in other form for the compensation of less prices as transferring money in their Jan Dhan accounts if high price of crop increase inflation. WTO and World Bank, has directed India to transfer

400 million people from agriculture to industries to fulfil the requirements of cheap labour for industrialisation and growth and this action of government has swallowed the agriculture sector. Now a days during and after lockdown the BJP ruled central government has bring changes as sale in free markets. Government is in a mood to abolish MSP and changes in Essential Commodity Act but if the government will allow the entry of private sector in agriculture for crop purchase and processing than the day will not be away when the commodity trading will start in India. Millers and processor will exploit the farmers after some time. We should never forget that India was able to win the race against COVID-19 only because of agriculture sector and if it has been in private hands than the black marketing of food grain will start. Now a days agrarian crisis has become social crisis.

Dual Policy of Central Government on Three Acts and Dilemma of Farmers.

Agitating farmers are enlightened farmers of Punjab, Haryana, Western U.P. and Rajasthan who can smells that govt. will rob Paul to pay Peter. Kisan Majdoor Sanghrash Samiti have distributed detailed pamphlets of long run harmful effects on agriculture and agriculturist of these three bills among farmers and labours who are partially or fully dependent on agriculture for their earnings. Everyone is speaking in same voice and their movement is organized and well planned. Farmers are in fear because, the way government imposed these bills during pandemic by calling a special session of Lok Sabha and passed these bills. So, the fear of farmers is natural. Govt. proved it itself when without any communication and meeting with farmers, in September imposed these bills on farmers and try to suppress their voice when Haryana govt. did laathicharge on farmers in Kurukshetra.

There are about 15 crore farmer families in India and out of these about 52% (NABARD REPORT) farmers are under debt in India. According to a study of WTO (2018-19) every farmer receive a subsidy of Rs.20,000 from all resources. Centre and state government together gives a subsidy of 3.25 lakh crore in the form of seeds, fertilizers, MSP,, electricity etc. but on the same time industries has been given a subsidy of 10 lakh crore and in America 45.22 lakh subsidy is given to each and every farmer. In1970 MSP of wheat was 76 rupees per quintal and today is rupees 1975 (26 times). Meanwhile, the income of the central govt. employees has increased 130 times and income of teachers has increased 320 to 380 times.

There is no relevance of the of trade and commerce bill in India as well as in Haryana and Punjab as number of marginal farmers is so large in India. About 86% percent farmers are marginal and small farmers with holdings below one hectares and 14% are semi-medium, medium and large farmers. In Haryana about 68% farmers are marginal and small farmers and in Punjab 33% are marginal and small farmers

Figure 1.

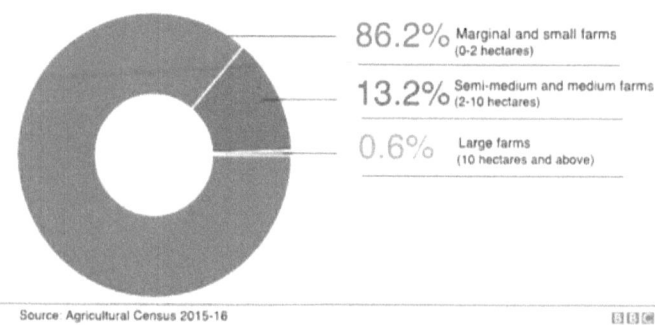

So the surplus production of wheat and rice per farmer is too low that they can not afford the cost of transportation to another state, moreover there is a socio economic relation between farmer and commission agent and we can say that commission agent is the ATM of the farmer from whom he can borrow money any time in needy hours and because of this social relationship the commission agent never reveal before any body about the debt of the farmer, as debt is considered a social stigma by the farmer community. I remember that my father used to borrow money from the commission agent at the time of sowing and harvesting of crops and at the time of any celebration in family. Shanta Kumar committee gives the excuse that only 6% farmers sell their product in APMC and small farmers still sell it to private traders. Govt. should increase the number of regulated markets so that every farmer can sell its products in regulated markets. We have only 7000 regulated markets in India and India need a vast network of regulated market in a radius of five kilometers and we need about 42000 regulated markets in India.

Figure. 2

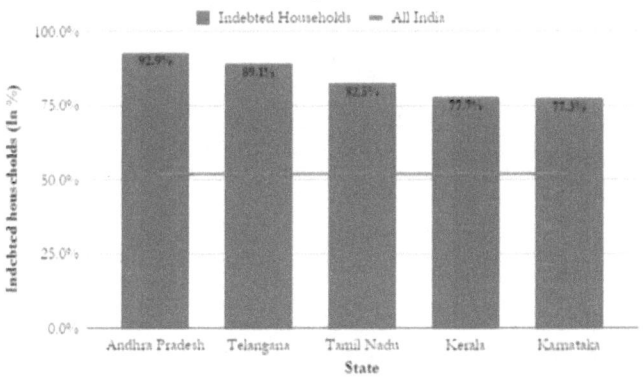

I listened these words from my childhood that, 'KISAN KARZ MEIN HI PAIDA HOTA HAI AUR KARZ MEIN HI MAR JATA HAI'. On an average each farmer has a debt of one lakh rupees and this debt is increasing in Geometric Progression and income of a farmer is frozen for the last 40 years. About 28 farmers commit suicide every day in India. On an average income of a famer family according to the report of NABARD is 8,931 rupees per month and it is less than the mobile recharge bill of a rich person. This all is because of the neglected attitude of each govt. towards agriculture sector. Farmers are fighting for MSP of their crops according to C2+50% formula suggested by Swaminathan committee. Prevailing MSP which focus on few crops, a few states and relatively a few farmers.In India we are thinking about a second green revolution where about 28 farmers commit suicide daily and 10,281 farmers commit suicide in 2019 according to NCRB data. On an average about 52% farmers are in debt in India.

MSP is must for agriculture products because of windfall losses in agriculture.Students of economics are familiar with the COBWEB theorem. A farmer takes decision to produce particular crop minimum six months before marketing so than a assurance of minimum support price is must in agriculture. Every industry can increase or decrease its production according to the demand in the market but a farmer can not as he has sown the crop six months before, it reach to market and standard law of demand and supply of Dr. Marshall do not work in food production. A farmer never knows as what will be market and weather conditions in coming next 5 to six months. Cobweb theorem describes this dilemma of a farmer. How the market forces and nature play a major role in deciding the price of agriculture products as two third of total land has no irrigation facilities. A farmer can be ruined because of lack of demand and low price. Even the plenty production can also ruin him when the price falls because of more production when supply surpasses demand and equilibrium price falls. This is called poverty in prosperity in economics. Even the developed capitalist economies, where free market supposedly r ule, highly subsidize its far ming communities. So the assurance of MSP is must for the crops. Govt gives MSP only for 23 crops and as I have written above that only 6% of total crop is sold on MSP because in loopholes in govt. marketing policies.The boom and bust of agriculture will be solved when we shift from state domination to individual freedom but govt. want to freeze the prices of agriculture products to increase its ratings to attract FDI and for cheap supply of raw material to industries. No doubt govt. want to set a new growth model but it should not be on the graves of farmers. About 46% population directly or indirectly depends on agriculture sector and even the vegetable vendors and fruit sellers they all will be spoiled as their earnings also depends on this sector.

Simple way to understand MSP existing and recommended by SWAMINATHAN COMMITTEE.

Cost A1 includes all expenses in cash and kind incurred by owner.

1. Value of hired human labour.

2. Value of owned machine labour.

3. Hired machinery charges.

4. Value of seed(both farm produce and purchased).

5. Value of pesticides and manure, owned and purchased value of fertilizers.

6. Irrigation charges.

7. Depriciation of farm implements.

8. Land revenue.

9. Interest on working capital.

Cost $A2 = A1 +$ rent paid for leased land.

Cost $B1 =$ Cost $A1 +$ Interest on value of owed capital assets (excluding land).

Cost $B2 =$ Cost $B1 +$ rental value of owned land and rent paid for leased- in land.

Cost C1 = Cost B1 + Imputed value of family labour.

Cost C2= Cost B2 + Imputed value of family labour.

According to Swaminathan report MSP should be = C2 + 50%. Existing formula for MSP is A2+FL

Contribution of Agriculture towards Self-Reliant India.

It is clear from the speech of our Prime Minister on 12th may and finance minister's press conference of 13th may that self-reliance is going to be a part of new economic policy and need of hour of Indian economy during this chronic time.chain of self- reliance will start from core of country (Agriculture sector) to district and to state. P.M as delivered in his speech that we have to live with this virus but have to *rejuvenate* our battered economy simultaneously. India's concept of self -reliance depends on five pillars quantum jump in economy, infrastructure, technology and demographic energy and creation of demand as well as supply chain in the economy. P.M announced a package of Rs.20 lakh crore (about 10% of GDP) as an economic booster for depressed and stagnant economy because of prolonged lockdown. In first trench Rs.3.7 lakh crore loan package for micro, small and medium enterprises (MSMEs) will be released and RBI announced a plan to infuse Rs.3.74 lakh crore liquidity while the finance minister came out with a Rs.1.73 lakh crore package for the economically weaker section. P.F(provident fund)contribution of employees has been reduced to 20 percent from 24 percent. For this package of Rs. 20 lakh crore government may borrow from banks or any other sources. Government can also go for deficit financing by printing new notes. No matter from where the money will come but it will increase the liquidity in the economy and liquidity will increase purchasing power and productivity in the economy. In 1930 when there was a great depression in Europe and all economist were unable to find solution of problem then J.M.Keynes a young economist suggested that government should adopt each and every method to increase purchasing power by deficit financing or by credit creation in the depressed economies. As a result, liquidity will increase in the economy and purchasing power will increase and people will demand for more goods and demand will increase, small industries will get inducement to invest more and supply of goods will increase in the markets. As in this package MSMEs have been given Rs.3.7 lakh crore for investment to fulfil the requirement of increased demand. This sector alone contribute about 30 % in GDP and gives employment to more than 12 crore people. But the most important thing will be that the demand for local products should increase instead of foreign products otherwise there will be outflow of money and there will be leakage in income generation process. Government can also give tax rebate to service sector to increase purchasing power in the economy. World famous economist and Nobel laureate Abhijit Banerjee also recommended that bottom 60 percent population should be given cash doles to spend to boost the economy. Because Marginal Propensity to Consume(MPC) of poor and middle class is greater than upper class and if there will be cash in their hands, they will spend a large portion of their income and as investment multiplier depends on MPC of the economy the income will increase many more times because of large multiplier effect.

As this package of Rs.20 lakh crore is going to be distributed in all sections of the economy as Prime minister said that farmers, small industries and labourers are the stakeholders of supply chain. Agriculture is the major stake holder of supply chain.Self- reliant agriculture will become when we will follow local to global.Our handicraft and jute products have a large demand in international markets, so these lost industries should be given a boost and promotion. India is a country of various geographical conditions and our products have demand in international market as basmati rice, turmeric, various spices and gur made from sugarcane. India can export these unique products in international market. Farmers should be financed by banks for this purpose and government should decide portfolio of loans to farmers in a transparent way for export promotion and for this change in crop diversification with a guarantee of MSP, so that the agriculture can be made a profession of profits and farmer an entrepreneur.

Perhaps Indian economy is moving on the path of balanced growth. All sectors should grow simultaneously as output of one sector is the input for another sector and service and agriculture sectors are the demand chain of industries and industries get raw materials from primary sector as Leontif proves in his input output model of interdependence of different sectors of the economy.

Let us hope for the best that this economic booster will prove as engine of growth and not remain only a survival dose. With self- reliant model we can overcome not only this pandemic time but post pandemic time also.

Crash of the Indian Economy but Agriculture Proved to be the Lifeline of the Economy.

Indian government made a squared planning of nationwide lockdown to defeat COVID-19 but the results of stay at home order became fully visible in the quarter1 of this financial year 2020-21. No doubt downturn and contraction in growth of the economy was expected by many estimating agencies but this cramp will be so drastic, was never estimated. Economy shrunk near to 24 % on the angle of 180 degrees in this quarter because of many exogenous and endogenous factors, a major one is severe lockdown of 70 days. India's GDP collapsed by 23.9% (in April to June Q1 data released by the National Statistics Office on 31st August) in this quarter and India's downfall was the worst among many major economies. Moreover, it is expected to come -7% in the whole financial year 2020-21. Contraction in GDP of 23.9% means that each and every Indian's income has reduced to 24% percent in comparison of Q4 of last financial year. There are no chances of recovery in coming 5 to 6 quarters according to experts. In other words, the total value of goods and services produced in India in April, May and June this year is 24% less than the total value of goods and services produced in India in the same three months last year

Almost all the major indicators of growth in the economy are at the lowest level of trade cycle. According to the data of Indian Ministry of Statistics and Program Implementation, all different sectors of the economy collapsed and only agriculture, which grew by 3.4%, in this quarter proved to be the lifeline of the economy.

Only Agriculture contributed to positive growth while Service and Industry contributed to the contraction in GDP (Figure). Agriculture is set to cushion the shock of the Covid-19 pandemic on the Indian economy in 2020- 21 with a growth of 3.4 per cent – resulting in an increase in its share in GDP to 19.9 per cent in 2020-21 from 17.8 per cent in 2019-20. This indicates that agricultural activities for rabi harvesting and kharif sowing were largely unaffected by the COVID induced lockdown. Industry and Services are estimated to contract by 9.6 per cent and 8.8 per cent during the year

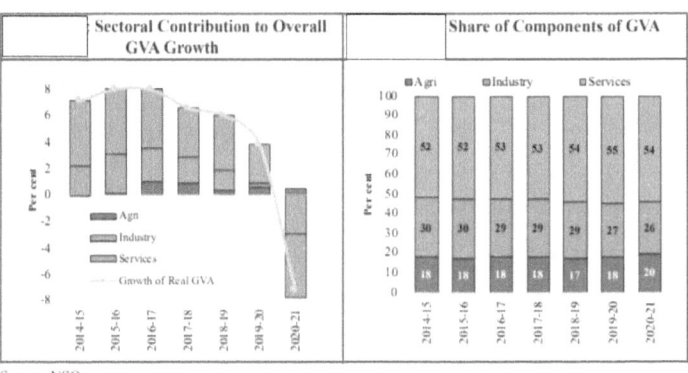

Source: NSO

The worst affected were construction (–50%), trade, hotels and other services (–47%), manufacturing (–39%), and mining (–23%). It is important to note that these are the sectors that create the maximum employment and value addition in the economy. In a scenario where each of these sectors is contracting so sharply, their output, incomes and investment are falling would lead to cyclical unemployment, not structural, having a hope of revival. Let's peep into sector wise downturn in AD or GDP contraction? Why hasn't the government been able to curb it?

GDP is the monetary value of the total number of final goods and services produced in an economy in a given financial year. Aggregate Demand is =C+I+G+NX or GDP.

The biggest component is consumption demand by consumers. This is C in the above given equation in the economy, this accounted for 56.4% of all GDP before this quarter. The second biggest component of the demand generated by investors. Let's call it I, and this accounted for 32% of all GDP in India.

The third component is the demand for goods and services by the government or the government expenditure (G) and it accounted for 11% of India's GDP before COVID time.

The last component is the net demand on GDP after we subtract imports from India's exports. These are net exports, inflow in the national income.

Consumption has fallen by 27%. In money terms, the fall is of Rs 5,31,803 crore over the same quarter last year because of decrease in the income of middle and lower class. Cash has become the king for the middle class because of financial uncertainty and the poor are penniless because they have lost their jobs during lockdown, so they have nothing to spend. Although the agriculture sector shows a positive growth but the increased income of the rural people can never be a substitute for urban demand.

Investments by businesses are more affected, it is half of what it was last year in the same quarter. In terms of money, the contract is Rs 5,33,003 crore. Investors has been given loan in loan mela to make fresh investment but because of lack of incentive of market consumer demand, they have not made fresh investment, either they have repaid debt or return it to RBI so the two biggest components (C+I) of AD which accounted for over 88% of total GDP, in Q1 were in worst condition.

Last component, government expenditure, went up by 16% but this was nowhere near enough to compensate for the loss of demand as government expenditure is mainly on health and to provide food security to the poor. When the demand from C and I fell by Rs 10,64,803 crore, the government's spending increased by just Rs 68,387 crore. Government's spending increased but it was so meagre that it could cover just 6% of the total fall in demand. According to Keynesian economics, when incomes fall sharply, private individuals cut back consumption. When private consumption decreases sharply, businesses stop investing. Since both of these are voluntary decisions in a democratic country, there is no way to force people to spend more and or investors to invest more in the current scenario until and unless government doesn't give any incentive to invest to investors and consumers to generate demand during depression. COVID-19 was an exogenous factor but nationwide lockdown was an endogenous factor and government can follow the strategy of regional containments instead of nationwide lockdown.

Under these circumstances, there is only one tool that can boost GDP and that is the government expenditure. Government can spend on social infrastructure either by building roads and bridges and paying salaries or by directly handing cash doles to the poor for the purpose of demand generation. If the government does not spend adequate amounts then the economy will take a long time to recover. Government can borrow from RBI and can sell its bonds or can-do deficit financing for the revival of the economy as we have solid infrastructure and the economy can revive soon. Centre and state government should make efforts together with a uniform strategy and economic policy and try to boost agriculture sector which has proven the lifeline of the economy and will boost the economy in near future.

Demand Generation by Middle Class and Agriculture sector to Tackle Depressed Economy.

Recently Rs.25000 crore flew back to RBI from commercial banks because MSMEs are not ready to take loan because of present market scenario of demand deficiency and are not in a mood to do investment as they are not finding it profitable. The root cause is declining purchasing power and consumption expenditure which is the major part of aggregate demand along with Investment and government expenditure. According to Credit Vidya (a survey agency) 52 percent of total workforce is without income or having income deficiency. It divides middle class market in three main segments. Affluent market (earning more than 60,000 per month) having 12 percent drop in income mid market (earning between 20,000- 60000) having 14 percent drop in income and mass market (earning 10,000-20,000) are having 56 percent drop in earnings. Mass, Mid and Affluent market's consumption is coming down by 54 percent, 24 percent and 26 percent respectively in this time. Mass market have cut its essential consumption but Affluent market and mid market are restricted to discretionary expenditure because of insecurity of jobs and distrust in economy and at this time cash is king for them. There can be some other factors which are responsible for lesser demand are social distancing and fear of contact. These middle classes (mass, mid and Affluent markets) are the 40 percent of population having maximum MPC and alone mass market comprises 70 percent of middle class market. Government has no plans for this section of population.

Poor class is on subsistence level and have nothing to spend. At the same time government has no money for public expenditure hence demand chain has distrupted and money is flowing back to RBI to which J.M.Keynes said about this situation in 1930 that money is in black box.FDI is also shrinking because of spread of virus and global economic policies. We are trapped in vicious circle of depression. Because of lockdown maximum industries has temporarily locked and it will take time to start production when demand will increase. We have to strengthen domestic supply chain sas well as global supply chain as all economies are interelated and for this, first of all we have to strengthen demand chain by increasing purchasing power but government is going towards supply side by giving loans to MSMEs without generating demand. No doubt revival of industries is must but inducement to investment will come from demand. When demand will increase of poor and middle class, it will generate demand chain to retailer, retailer to wholesaler and finally wholesaler to industries. It will give incentive to investment and will increase productivity and supply and as we have already a solid infrastructure and capacity to produce and supply will increase in its slanting nature.

The most effective weapon is increasing purchasing power of middle and poor class as they capture a big proportion of market. Easy and cheap loans can be provided to middle class for consumer durables as for vehicles and household appliances, side by side increase disposable income by lesser tax deductions. Agriculture sector is growing at more than 5 percent, so there are more possibilities of demand generation. Government can increase the income of farmers by giving them appropriate MSP so the purchasing power of farmers can increase, moreover subsidies can be given on agricultural implements and when they will demand implements it will give a shock booster to production in capital goods industries.

Direct Benefit Transfers (DBT) with benefits as PDS and subsidies on essential household needs as on fuel and electricity can be given to poor section to increase their purchasing power. Although government try to infuse money to poor by MGNREGA and ration to eat per family but this is not sufficient. As Abhijeet Banerjee says 'print new notes' for DBT. No doubt government fiscal deficit and inflation rate will increase by giving DBT but after some time all will be automatically settled when production will increase, moreover our focus should be on short run remedies.According to NSSO in this chronic period of Covid -19 on anaverage Monthly Per Capita Consumption Expenditure (MPCE) has decreased 8.33 percent of every individual and if we calculate for these three months it is about 25 percent less and if compensatory DBTcan be given to poor than most persons can come back to their pre.covid levels of MPCE.

The worse time is ahead Poverty inequality both will increase.

To increase the level of output, one has to enhance the demand in the system. Indian economy will take time to get revive as all sectors of economy are in depression but will revive soon because we have a solid infrastructure and with increased demand temporary bottlenecks in supply chain will disappear. Lastly, we can not isolate our economy in this age of global world rather we should compete with other economies including China with quality products for smooth functioning of supply chain.

MSP is must as a Safety Shield for Farmers.

MSP is a crucial and key issue for the survival of agriculture as it works as a social safety net for farmers. The unpredictability of agricultural yields and products often results in windfall losses making the lives of farmers debt-ridden and pathetic. The economic decisions involved in agriculture are massive and need to be recognized as such. The demand of Minimum Support Price (MSP) is directly linked with food security in India. Hence, farmers have a rationale and urgency in seeking govt attention towards their economic well-being as made or marred by new agricultural legislations of the central govt. For instance, students of economics are familiar with the COBWEB theorem. A farmer have to take decision to produce particular crop minimum six months before marketing so then a assurance of minimum support price is must in agriculture. Every industry can increase or decrease its production according to the demand in the market but a farmer can not as he has sown the crop six months before, it reach to market and standard law of demand and supply of Dr. Marshall does not work in food production. A farmer never knows as what will be market and weather conditions in coming next 5 to six months. Cobweb theorem describes this dilemma of a farmer. How the market forces and nature play a major role in deciding the price of agriculture products as two third of total land has no irrigation facilities in our country. A farmer can be ruined because of lack of demand and low price. Even the plenty production can also ruin him when the price falls because of more production when supply surpasses demand and equilibrium price falls. This is called poverty in prosperity in economics. Even the developed capitalist economies, where free market supposedly rule, highly subsidize its farming communities. So the assurance of MSP is must for the crops. Govt gives MSP only for 23 crops and only 6% of total crop is sold on MSP because of loopholes in govt. marketing policies. The boom and bust of agriculture will be solved when we shift from state domination to individual freedom but government want to freeze the prices of agriculture products to increase the ratings to attract more FDI and for cheap supply of raw material to industries. No doubt government want to set a new growth model but it should not be on the graves of farmers. About 46% population directly or indirectly depends on agriculture sector and even the vegetable vendors and fruit sellers they all will be spoiled as their earnings also depends on this sector.

According to Swaminathan committee the MSP should be C2+50%.there is a difference between existing MSP calculated with reference cost A2+ FL and that calculated with reference cost C2 for the main crops. In other words, the promised margin of 50% over A2+ FL the current basis for MSP hardly brings about an increment in farmers income in actual. Therefore, Swaminathan's proposed MSP formulaC2 plus 50%is relatively a better approach towards providing farmers deserved value for their produce. However, it is proposed that to strengthen the financial well-being of farmers and to double their income the reference cost should be taken as C2 including all CoP rented and imputed.The imputed cost of the labour of head of the family should be counted as an entrepreneur's cost.

According to Swaminathan report, the MSP, suggestions (C2+ 50%) for wheat and paddy, the MSP for wheat would be `2,787 and paddy `3,116, which is higher than the existing MSP by about 45% and 67%, respectively. One of the key demands of the ongoing farmers' agitation is that a legal provision of MSP ismust and purchase below MSP should be punishable. The reason for this demand is that even though the MSP is declared for 23 crops, most of them are usually sold at much lower rates. It is mainly two crops – paddy and wheat – that actually sell at or above MSP and even this is usually found to be the case in Punjab and Haryana. And that is also why farmers of these two states stand to lose the most if indeed procurement at the MSP is curtailed.

That the announcement of MSP for most crops remains mostly on paper and in the absence of effective procurement which is at its best in Haryana and Punjab farmers are rarely able to demand prices which are at par or above MSP.

For instance, the paddy having MSP Rs.1868 was sold on Rs.1100per quintal,maize having MSP 1850 was sold at Rs. 800 and cotton having declared MSP Rs. 5515 was sold on market price of 3200 per quintal in October and November.

Because ofIncome denied due to sales below MSP in kharif 2020 of all crops, national loss in October was 1099 crore and in November it was 793 crore (Source:AGMARKNET).

The only states with net gains over the two month period were Haryana and Punjab, however, even in Punjab, farmers have to sell maize and cotton below the MSP.

If we compare India with other developing countries in context of subsidy given to farmers, the fresh report of OECD (2020-2029) is very surprising.The countries of Asian pacific (India also fall in this category) region gives more subsidy to farmers and agriculture than India. China gives 10% more prices to its farmers in comparison of international market but in India farmers get 13% less value of their product, if we compare it with international market. Prices of agriculture products are curbed intentionally as inflation was always a political matter. Besides the cost of production, many economic and non-economic factors at both national and international levels play an important role in the determination of MSP. Althoughit is evident from the report of global hunger index that we have not sufficient food security but still we are careless about this sector of the economy.

OECD country report on India states that support to farmers is negative net support of -5.0% in 2017-19

"The negative value of the PSE(Producer support estimates) reflects that domestic producers, overall, continue to be implicitly taxed, as budgetary payments to farmers do not offset the price- depressing effect of complex domestic regulations and trade policy measures,".

Analysing the farm price-depressing effect on producers, the policies provide implicit support to consumers. Policies that affect farm prices, along with food subsidies under the Targeted Public Distribution System, reduced consumption expenditure by 21.4% (%CSE) on average across all commodities in 2017-19." If we apply cost benefit analysis than we find that the consumer's benefit is on the cost of farmers.

In other words, contrary to the perception among many, the government has preferred the welfare of Indian consumers over the Indian farmers.

Moreover government want to escape from the purchase on MSP as it's total expenditure is about ¾ of total revenue if govt. purchase all 23 crops on MSP. These new farm laws restrict state interference and by introducing these laws central government supportlessize faire policy of demand and supply forces and free market. Farmers have apprehension that they will be in the clutches merciless private purchasers and the current scenario is mirroring their fear.

Reverse Migration and Preparation of Source State for this unemployed workforce

The ultimate source in economic development are people, not capital or raw material

— Peter Drucker

A severe lockdown was announced on 24 March, 2020 and 'stay home stay safe ' became the slogan to flatten the curve of pandemic in India and it continuous up to 4.0 with some relaxation and with some rules and regulations in containment zones but the graph of pandemic continued to rise despite the restrictions. Since the beginning of the nation wide lockdown due to covid-19,India and the world have seen on media, the illegal and unsafe movements of barefooted migrants and miserable and disturbing pictures of thousands of migrant workers have shattered the people not only in India but all around the world. Many of them lost their lives in trains and on roads.The capacity to 'stay home and stay safe' of about 10 crore of India's population has no meaning because they neither lived in what could be called a home,nor were they safe inside their dwellings given the accepted norms of social distancing and hygiene as we saw in Dharavi.This invisible force (migrant labourers) who create economic surplus in state in which they work but having no political role as they have no vote in destination state, so that the government of that state doesn't have any labour commission for them and they are totally ignored.They were invisibles for us until we saw them on roads going towards source state. This invisible force provide support services to each and every sector of the economy and across the classes. Government has declared pandemic as natural disaster and made obligatory for the employees to pay the wages, as per National Disaster Management act but it is not feasible, moreover difficult to assess it's implementation because of many reasons, firstly administration have no record of them and second major reason is that the income earned by the self employed migratory workers can not be termed as wages and they have to remain without earnings in absence of any work. They are rikshapullers,road side teastallers,cobblers, pressers and doing small businesses as painter, carpenter and many more. And even in their 'destination' as MSMEs, tiny industries and agriculture the financial conditions are too fragile to release wages in absence of any production. Only the domestic workers may have received some money to survive themselves. When the lockdown was announced nothing was thought about these labourers. The word pandemic was so dreadful and it triggered massive reverse migration from 'destination' to 'source' in large part of the country.Although the government has provided them ration and money but now return to home becomes their psychological problem. Government get trapped in double disaster one was of pandemic and second was self generated that was of migrants.

According to professor Amitabh Kundu in an estimation of 2020 there are 65 crore internal migration and 33% of them work in informal sector as labourers. Prof.Kundu's estimation shows that Uttar Pardesh and Bihar account for the origin of 25% and 14% of the interstate migration. 40 to 60 lakh people would be wanting to return U.P and 6 to 9 lakh to M.P Majority of migrants were marginal farmers in their source state and because of debt, low productivity, water scarcity, floods and crisis of other livelihoods options have forced them to migrate in destination state. As Lewis (1950) and Harris Todaro says in their models that the labour migrate from subsistence sector to developed sector in a dual economy. According to Lewis this surplus labour whom productivity is zero in subsistence sector can create surplus in developed sector. Todaro says that when the expected wage rate in developed sector is higher thansubsistence wage rate in primary sector, it will induce migration. Although it is nearly impossible to measure the size of this work force in the urban informal sector, but when reverse migration began due to pandemic, there was an opportunity to prepare a data base of this labour market, which is important as given contribution of them in economic surplus and GDP.Counting and mapping of skill of this huge army would have been possible, if for example, the registration form which they have to take to return home could include the details of skill, nature of employment, income earned and many more information, as if they would like to return in near future. Inadequate information may be a strong hurdle in proper utilisation of funds allotted for help of migrant workers. This information may be useful for both the states of destination and state of source, especially if a sizeable portion of this population does not wish to return. In post covid time there are chances of surplus labour in some states and relatively deficits in others.The administration than can frame a better picture of the need for employment generation programs in surplus states.New MSMEs can be established in labour surplus states and their skill may be used. As Uttar Pardesh govt.has done in skill mapping of migrant workers.According to U.P government official data about 23-24 lakh migrant workers have returned to U.P during the lockdown. In their skill mapping it was found that they are large number of painters, carpenters, drivers,electricians and persons associated with electronics,security guard,furniture and fitting workers and auto repair mechanics. If employees want them to come back they have to offer better working and living conditions to attract workers as by giving them free accommodation, free education for their childrens and medical facilities.It is also quite possible that an introduction of capital intensive technology can replace this labour in destination states in near future. The source state government can collaborate with ILO to solve the problem of migrants and state governments have to improve infrastructure with establishment of many more SMSEs to absorb this labour. As Nurkse says that this unlimited supply of labour in LDC is a potential source to increase productivity. Now we are in a situation of reverse migration and this is the proper time to make use of the skill of return migrants and make them a source of economic growth by giving them work according to their skill, as government of U.P is trying to do and if they gainfully employed than they can become an engine of economic growth in' source' states and if this problem is not handled timely it will have dangerous spill over effects on society.

Can India Become a World Vendor: A COVID Time Opportunism

On 15th August speech Prime Minister focused on value addition He announced that we will not merely remain the supplier of raw material but will supply final goods to world. We will provide best infrastructure and skilled labour to MNCs. Manufacturing and value addition are the mantra of development. FDI, especially in manufacturing, tends to boost growth, moreover it is also helpful in technological transfers.

But sorry to say that P.M's dream to make India a global factory and world vendor is a herculean task for India because of many reasons like insufficient expenditure on R&D, flexible labour laws, democratic system, lack of skilled labour, insufficient solid infrasture and many more as the list is too long. No doubt India is lagging behind China but after COVID-19 it is favourable opportunity for India as world economies doesn't want to put all manufacturing eggs in one basket and second choice may be India as India has a big perportion of young population and MNCs need market for their product and India can take it's demographic advantage, but we have to provide a better business environment to MNCs. R&D is equally important for innovations and for the purpose to minimize cost and to save time, side by side for skill and quality products. So the educational institutes have a pivotal role in innovations and advance technology. Expenditure on education should be increased for the purpose of skilled labour and for fresh discoveries. China has filed more applications for patent in comparison of all other countries and even US.

India is not the sole country who tries to start new love affair with MNCs, as there are others -Korea, Taiwan,Vietnam,

Thailand and ofcourse China and these all are in the priority list of MNCs. According to study of a Japanese financial group (Normura) in 2018, 56 companies relocated their production base from China, of these 26 went to Vietnam, 11 toTaiwan and 8 to Thailand. Only three came to India.

Modi's mission of self-reliant, permotion of local brands and boycott of Chinese products may have reverse impact upon our industries.We very well knows that political and economic matters are intermingled and China's coercive policy must have negative impact on our trade with China but China is deeply embedded in not only in Indian economy, but in global economy also. So containment of China is not possible but we can put some constraints on China and side by side should try to be aligned with China. Although economic matters never stand in isolation because of political matters, in the current year as the relations with China has deteriorated.The need to free the country from Chinese imports may increase the cost of domestic final product because for many intermediate products we depends on China and their cheap substitute is not available in our country.

Arvind Pangariya (former vice chairman of NITI Aayog) said," I am worried about rising tarrifs, which will be hurdle in the growth of firms,".There is misconception that local products are economic substitute of imports from China, and there is every chance that the principle of trade advantage will be violated and welfare of the world economies will also be violated. Because of overprotectionist policies of the government can reduce FDI in India.

Focus should be on minimizing the production cost by importing comparatively cheap inputs for production. Although officially denied, it is of concern that 80 percent of global supply of crucial pharmaceutical ingredients is made in China.China makes 70 % of acetaminophen used in America. India depends on China for 80% of its supply of active pharmaceutical ingredients (API). Overall, China makes half of the planet's API, according to Britain's Medicine and Healthcare Products Regulatory Agency and pharmaceutical analysts.How long India will take to produce these through Atmanirbharta?. Producer are worried about the cheap inputs and intermediate good for production process if government put restriction on imports from China.

According to chairman of Maruti Suzuki that the components of its cars are imported from China and its economic substitute can not be produced in India and if produced than they are costly and not of right quality, so the economic burden is going to be shifted to the pocket of Indian consumers. Same is the case of defence ministry as the burden is going to shifted on tax payers or the deficit finance will increase.

These actions of government will take Indian manufacturing away from global competitiveness and the chances of Indian brands making it global brands will go down.We should follow the policy of Deng(Chinese leader) that our ultimate goal should be development, that either the cat is black or white, main purpose is to catch the rats and this policy of Dang in 1979 made China a global factory by attracting MNCs in China. No doubt China's wants hierarchy of nations, but it's the right time for India to take an opportunity of COVID time and the bitterness in the relations of US and China and try to align India-US as well as India-China trade relations (the biggest and cheapest global supply factory of world) for disinterupted supply chain of intermediate goods for our manufacturing sector, so that it can do further value addition. As for as US is concerned, we export 28% of total medicines to US and about every third tablet used in US is exported from India, but Trump is going to change its import policies, so Trump's America is not a reliable trade partner for India. This is the most favourable time for India to promote FDI and to be the vendor of world economies. India should try to follow the policy of neutrality in Quad also, recently established by four countries to control China's challenges and territorial ambitions in south china sea and the Indian ocean. Our focus should be on FDI and to provide a better business environment and better infrasture to attract MNCs.

Cost Benefits analysis of Trade with China 'The Global Vendor'

Earlier, India and China engaged in deadly clash, the worst in 45 years. A national anger over the killing of Indian soldiers has nationwide protest to boycott of Chinese goods. Although it is not easy to forget the sacrifice of the lives of our soldiers but without finding any other cost wise cheap substitutes of Chinese imports specially of intermediate goods, banning of Chinese imports may ruin our industries and there is every possibility that wrong decisions can be taken in external trade matters by the government. India may import war weapons and also going to impose trade barriers, tariffs, import duties, order cancellations and bans on imports from China. Revenge is a dish best served cold and one thing should be clear that India will loose more than China if we will flow in emotions. We should avoid emotional pressure on government to boycott of Chinese goods. The trade with China has reduced over the past two months due to pandemic but after deadly clash in Galwan valley it will have worse effects. No doubt dragon always indulge in cunning activities and always play with economic and political weapons simultaneously as it did with Japan and Seoul in 2017 when it's citizens boycotted and vandalised Toyota and Honda showrooms. In reaction of attack, Indian railway just cancelled the Rs.471 crore contract awarded to a Bijjing firm, state owned telco BSNL has been instructed not to use gear from China's Huawei for network upgrade, Vivo's Rs.2199 crore sponsorship of IPL is in doldrums and under review. On june 23 the Confederation of All India Traders released a list of 500 categories of products imported from China that can be substituted with goods made in China.

Union minister Nitin Gadkari is supporting the policy of import substitution and advocates local brands in a webinar, and RamadasAthawala goes one step ahead and demanding ban on restaurants selling Chinese foods (made by Indians). Without analysing the cost benefits of these trade barriers and bans on Indian economy may deteriorate our budget. Trade theories explore that we do trade for our benefits, not for others.As Adam Smith says that" baker, baked the bread not for us but for his own benefit". Japan who is enemy of China and still do 7% of it's total trade with China but India does only 2.1% of it's total trade with China. If we think that China will start shivering because of India's ban on it's imports, this is our misconception and myth. China has become global factory and has grasped global supply chain and India is not alone who depends on China for intermediate goods, semi manufactured goods and raw material. From 1970 China adopting the policy of maximising it's economic benefits, as Dang said in late 1970s that either the cat is black or white, our only purpose is to catch the rats. China is Asia's largest economy and world's second biggest economy with a GDP $13.6 trillion because of it's political structure, production strategy, population trends, and solid and huge infrastructure. Government of China provide best infrastructure and facilities to producers and it's efficiency to minimize production cost has make it global factory and as a result dominate global value and supply chain for many key industries of all world that make China global leader.Production is the by product of all above mentioned facilities. The whole world depends on China from cars to phone, train to aeroplane and garments to laptop.

There are several reasons for China's dominance in global supply chain.China's school education policy from 1970 taught theory and practice in combined curriculum. Chinese labour knows how to produce goods that are not just cheap and durable but also fulfil the demands of different segments of different economies. About 80 % labour in china have basic education and skill and in India only 25 to 30% labour has basic school education. China's primary sector have capacity to produce and process, with out any barriers and hurdles and the final production reach to final consumer. But in India where is rural industries and how we can achieve the mission of Atamnirbhar Bharat.Both India and China roughly have the same population, but China's economy is five times larger than India.GDP of India and China was almost same in 1985 but after it China's GDP per capita increased to $8,827 and India's $1,942.Both have made enormous growth but China has done miracles.

In 1970s economic reforms were introduced in China but in India reforms came late in 1991. China has communist system and a centralized controlled economy and when government of China takes any decision, there are few hurdle in it's way and we have democracy and decentralized system, although this system is inclusive but development becomes nightmare in this democratic system.

In 1980 China announced the policy of single child and in coming years perportion of it's working population (15-64 years) comes down and China changed it's strategy simultaneously, as it started producing those goods in which value added per worker was more and adopted export led policy for growth. China do skill based work where per worker value added is multiple times greater in comparison of India. China's value added per worker in manufacturing is the biggest evidence of it's growing technological powers.It has built a huge research and development ecosystem within it's country as clear from it's patent applications, as their number are more than all other countries even U.S.A. hence it is clear that India is not in a position to compete with China.India has also a large proportion of young population in it's total population but our population becomes liability instead of asset because of lack of infrastructure and job oriented opportunities and corruption. It is easy to delete Chinese app from the mobile but boycott of Chinese goods is not so easy as supply chain of major industries depends on imports from China.Any way we can boycott consumption goods but expenditure of consumption goods is just 20% of total import bill.

India's pharmaceutical industries import about 68% Active pharmaceutical ingredients (API) from China and around 80% of all solar equipments are also imported from China. In Mobile phone, China capture 74% of total market leaving Samsung and Apple behind.If India will impose restrictions than producer will be the sufferer and it will have negative impact on growth rate in India.A mind blowing words said by someone came in my mind that in 1949 there was a question how China will survive and answer was that socialism will save China and after the disintegration of U.S.S.R, question arises that who will save socialism, it was said that China will save socialism.Now Socialism will save capitalism as China is the global supply factory as well as a huge market for capitalist countries.

Economic Inequality: The Other Side Of Pandemic

Ever since India got Independence, the country has gone through a wave of ups and downs to finally become a global powerhouse over the last five years. India's growth story from beginning is of a struggling economy in the 1950s to a $2.9 trillion economy in 2020 may be impressive, but not enough to support its continuously growing population.

This is why India is one of the major economies in the world having unequal distribution of wealth. Even after all these years, there has been no change in this trend as the rich tends to get richer and the poor, poorer.

The Covid-19 crisis has only made matters worse for India's middle and low income groups and difficult to revive as the pandemic may leave a permanent scar on the economy. 2 crore people in organised and 10 crore in unorganised sector lost their employment during pandemic which further will add to more inequalities.

The popular belief that coronavirus pandemic has made everyone poorer is not entirely true.While most economic activities are suffering due to restrictions imposed during the lockdown, some of the world's top billionaires have seen their wealth rise. Ambani who is the sixth richest man in the world, has added $30.5 billion to his wealth this year.Cyrus Poonawala, the founder of Serum institute of India, saw his wealth grow sharply this year as well. At the same time, the world's poor have seen their wealth deplete.

In India, the wealth gap has been rising sharply during the ongoing pandemic as previous levels of income inequality were already high in the country.

India's GDP story is very ridiculous as according to data we crossed UK two years ago, France last Year and will cross Germany and Japan in the next five years. That will leave only America and China ahead of us. But our per capita GDP and inequality story is embarrassing and COVID-19 lockdown proved that how per capita GDP and equal distribution of wealth is more important for our economy than total GDP. The income of an industrialist increases billion time per hour. What a contradiction ?, that we are living in a economy where GDP increases but we are loosing the rank of HDI year by year. That's the reason we are standing on 94th position out of 107 countries in Global Hunger Index. In 2000 the upper 1% people were having 37% of total wealth of the economy, In 2005 it increases upto 43 %, 2010 increase 48%, 2014 increase 58% and in 2020 it is 62%. It clearly indicates that remaining 99% have no meaning of growth. In coming years this inequality will increase because of COVID pause according to IMF.

73% of the wealth generated in 2017 went to the richest 1%, while 67 million Indians who comprise the poorest half of the population saw only 1% increase in their wealth. There are 119 Billionaires in India. There number increased from 9 in 2000 to 101 in 2017.Between 2012 and 2018, India is estimated to produce 70 new millionaires every year. The question arises, is Make In India Abhiyaan making millionaires? In India, the wealth gap has been rising sharply during the pandemic as previous level of income inequality were already high in the economy. 23crore people in India sleep hungry every day.

In January 2020 study by rights group Oxfam India suggest that India's richest 1% hold more than four times the wealth held by the 953 million people who make up for the bottom 70% of the country population, overtaking the British Raj's record of the share of the top 1% in national income, which was 20.7% in 1939–40.The study added that India's top ten percent of the population holds over 74% of the total national wealth.

Commenting on the study, Oxfam India CEO Amitabh Behar had said: "The gap between rich and poor can't be resolved without inequality busting policies, and too few government are committed to these."

"Our broken economies are lining the pockets of billionaires and big business at the expense of ordinary men and women. No wonder people are starting to question whether billionaires should even exist."

Since the study was from January 2020, it does not capture the economic devastation caused by the coronavirus pandemic. An International Labour Organisation (ILO) report, which predicts that 40 crore Indians may be pushed in poverty, offers a better picture of the widening wealth gap in India.

And government cannot control it without opening the strings of it's purse in this chronic time.

Let's try to peep into the poverty and inequality in India with the help of Gini coefficient.The Gini coefficient, sometimes called the Gini Index or Gini ratio, is a statistical measure of distribution intended to represent the income or wealth distribution of a nation. Gini index < 0.2 represents perfect income equality, 0.2–0.3 relative equality, 0.3–0.4 adequate equality, 0.4–0.5 big income gap, and above 0.5 represents severe income gap.

India's Gini coefficient is 0.832, Pakistan's is 0.67 and Nepal's is 0.71. The worst Gini coefficient is of the most populous countries like China and India, having more than 70% of the world population. The UnitedStates has a Gini coefficient of 0.480. In 1990, the Gini coefficient was 0.43, indicating an overall increase in income inequality over the last 30 years. This coefficient is rising with increasing GDP in all economies.This is because of the economic growth model of all the economies including India. While the Gini coefficient is a useful tool for analyzing the wealth or income distribution in a country, it should not be used as an indicator of a country's overall wealth or income.

Income inequality has both political and economic impacts such as slower GDP growth, reduced income mobility, greater household debt, political polarization, and higher poverty rates. We should never forget that poverty anywhere is a threat to prosperity everywhere.

The Unfolded Story of Health Sector: When Wealth Decides Health.

The Novel Corona virus has generated a global public health crisis and impacted all the countries irrespective of their development status. During pandemic we all observe that the leaders and bureaucrats preferred to go in private hospitals like Medanta instead of government hospitals for the treatment of COVID. Our leaders have no faith in government hospitals and services of staff of government hospitals. This all unfolds the story of health sector as well as dual character of our leaders. Because of this impression common man too have no faith in government hospitals.

Let's try to analyse health sector of India. India's public health expenditure has been rising over the last decade in order to combat the health issues of its growing population but not an expected rate. It was estimated to be around '1.28 percent of the countrysGDP and this figure is almost constant for the last 18 years.In 2000-2001 it was1.3 % and in 2017 and 2018 it shows a nominal increase of 0.1%.Including the private sector, the total healthcare spending in the country rose to 3.6 per cent of GDP in 2016 but this is very low when we compare with other countries. The average for OECD countries in 2018 was 8.8 per cent of GDP while the healthcare expenditure in the developed countries like the US was 16.9 per cent, China was 5 per cent, Germany was 11.2 per cent, France was 11.2 per cent and Japan was 10.9 per cent, according to a report of Care Ratings. The report added that the per capita total healthcare spending in India was $209 in 2016 as against the OECD average of $3,994 in 2018.The severity of the situation can be better understood by the fact that India ranks 170 out of 188 countries in domestic general government health expenditure as a percentage of GDP, as per the Global Health Expenditure database 2016 of WHO.

According to world bank report India spends (1.2%) equivalent to least developed countries even less than Sri Lanka (1.6%) of their GDP where world average is 5.9%. India's expenditure on health is less than the group of the countries (Middle Income Countries) to which India belongs. MIC spends on an average 3% of their GDP on health.. Indian government spend about 90 % of it's total expenditure for productive purpose so the health sector has been ignored asIn developing economies health of population is not on priority. The result of the low level of investment on health by government sector has resulted a huge expenditure by private sector (more than 50 %). According to National Health Account, Indian people's out of pocket expenditure is about 63%of total health expenditure. Out of total expenditure on health in India (3.6%) only 1.2 % is done by public sector and rest is by private sector. State government incur about 70 % of health expenditure.

The government aims to raise the public health expenditure to 2.5 per cent of GDP by 2025 in a time-bound manner but this expenditure is not sufficient as it should be minimum 4 to 5 % of GDP. However, the expenditure budget on medical research has not increased significantly over the past few years. The expenditure budget for research was Rs 1727 crore in FY19, which has grown to Rs 2100 crore in FY21, according to the Budget documents.Poor and middleclass people are forced to go to private providers, although they are unable to afford huge out of pocket expenditure. Finance plays a big role in investment in not only in health related expenditures but in medical research, education, equipment and technology also.

life of each and every person is precious and any game with the life of every Indian should be unpardonable offence.

According to a study based on National Sample Survey Organisation reveals that India has 20.6 health workers per 10,000 people. While it is less than the World health Organisation's minimum threshold of 22.8.There is minor increase in numbers from 19 health workers per 10,000 people in 2012. The study also reports a disparity in the density of doctors and nurses across the country. The number of doctors in Kerala and UTs is high as compared to larger states such as Rajasthan, Jharkhand and Bihar.Unfortunately, the distribution of health workers is uneven not only in states butbetween urban and rural areas also. Rural areas having nearly 71% of India's population and have only 36% of health workers.

Overall, 'Indias per capita expenditure on health is about 1,600 rupees(include public and out of pocket in 2018). However, this was national average estimate with some parts of the country focusing on health far more than others. For example, the larger and wealthier states spent 120 to 180 billion rupees on health in fiscal year 2018; whereas north-eastern states which are poorer and more remote, spent around three to six billion rupees on healthcare that year.This is a situation when wealth decides health. A poor person becomemorepoor because of this out of pocket expenditure (private expenditure) on health. In economics we call this vicious circle of poverty and because of this circle,a poor always remains poor. According to an estimate about 6 percent people becomes poor because of health expenditure but this figure seems to be too short as this number seems be much large.

The low health expenditure by the government has led to a highly developed private health care sector. Private hospitals make up more than half of the country's healthcare infrastructure but the health preference of private and public sector are different. Aim of private health expenditure is on multispeciality hospitals and but in overpopulated countries like India we need Primary Health Centres for awareness about health and nutrition so that preventive measures can be adopted by the public. About 70 to 80 percent health services are provided at PHC. Preventive measures of illness are important part of health as large number of diseases can be controlled at initial stage. Less expenditure on PHC in India leads to an unbreakable chain of communicable diseases.

However, this is not to say that the government is not working towards its goal of healthcare mission. Various programs like the Ayushman Bharat and the National Health Mission have already showed some success by providing the common man with an alternative to exorbitant healthcare costs and treatments but the major drawback of this programme is that this demands secret data of beneficiaries.

Because of this novel coronavirus pandemic crisis, the government has increased the spending on health care but it has become necessary to increase this spending not only to meet the ongoing health crisis but also to alleviate any future pandemic conditions as health is the engine of growth and a healthy person is an asset for the economy. IMF has said in its annual Article IV reports that India can boost its human capital's productivity by investing in education and healthcare and nutrition. In 2018, it identified poor public health as the most important hurdle in development. One thing we all observed that immunity played a big role against this battle of COVID 19. According to a survey upper 20 percent people are more affected by this virus. Poor migrants and villagers have strong immunity, that's the reason of less mortality rate in rural and poor India because of this virus. A health slogan of ' food should be medicine and medicine should be food ' should be started in near future.

Need of a Paradigm Shift in Economics

In 1991 we adopted the policy of LPG and removed restriction on international trade and opened our doors and windows for international trade and for FDI by making changes in multilateral and bilateral trade policies. But now the central government want to promote the growth model of ATMANIRBHARTA and SELF RELIENT, with a slogan of Be vocal for local and local to global to heal the battered economy.

As India is member of WTO and want to attract more FDI to fulfill its dream to become global vendor of manufacturing goods. Central government's efforts to improve the ratings of international agencies to attract more FDI and want to capture the place of China of global factory, but on the other side with a slogan of Atam-Nirbhar Bharat Abhiyan, where we are going…? If we will impose trade restrictions to protect our manufacturing industries and will promote only domestic products and unfortunately if all economies also followed the policy of import substitution and restriction then where we will sell our exports. I always find myself in a dilemma whenever I try to understand the current international trade policies of India.

A 2.9 trillion dollar(current GDP) economy is going to set its national income limits by imposing trade restrictions, as income from abroad is a major part of National income of a developing economy like India. Make in India was started with a target to increase the participation of manufacturing sector in GDP upto 25% which was 13.7% in 2019 and was much better(15.6%)in 2015, when the movement of make in India was started. Make in India movement opened the doors for multinationals to invest in India but now when we focus on vocal for local, it means we want to manufacture everything without foreign assistance and foreign investment. No doubt it can increase employment rate but FDI by multinationals will be demoralize.

Our main motive should be to achieve previous level of growth with equal distribution of wealth. In India the wealth gap has been rising sharply during the ongoing pandemic time as previous level of income inequality were already high.Fiscal deficit is increasing day by day, GDP and tax ratio is decreasing and all macro variables are in dwindling state. No- productive expenditures of government are increasing.

Consumption is decreasing and restricted to consumption of essentials. Consumption of discretionary items will be postponed and will fall drastically.

Due to shortage of resources with the government, investment is declining as maximum resources are diverted to heath expenditure. Due to huge unutilised and underutilised capacity and lack of demand, new investment by private sector will be low during lockdown and post-lockdown period.

As incomes of all sections drop, savings will fall. Only those sectors that continue to function will show some savings. There will be negative savings in the economy. Businesses that were closed will use reserves and will dissave and Banks, the creditors for businesses, will dissave since they will pay interest to their depositors but will not receive the interest due to moratorium on repayment. Workers and migratory labour have lost their work, have no income but will continue to consume, so, they will dissave. So there are no savings in the economy. Even the middle class has cut their consumption and cash has become king for them. Agriculture sector have savings but their MPC is always low as they have little habit of spending.

Revenue of Government (Tax Revenue + Non-tax Revenue) is falling continuously.

T will fall substantially as salaried class have lost their jobs. Indirect tax collection will drastically fall as production declines and especially of luxuries, that are in the higher tax brackets. Most production will consist of the essentials and they pay either zero tax or a low tax rate.

So, the tax/GDP ratio has fallen. In India, it come down from 16% in 2019 to 8% in present time. Since both GDP and tax/GDP ratio will fall, revenue will fall sharply. Non-tax revenue will also fall. In India, it will be substantial.

Expenditures are on salaries, public investment, purchases and transfers, like subsidies, interest payment to borrowers and now to the unemployed for their survival. Expenditure of the govt. will increase continuously as interest payment and salaries are committed expenditure. Defence expenditure is treated as a holy cow, so it is difficult to curtail. Social sector expenditure and public investment are usually cut. Subsidies and will rise substantially to help the unemployed and poor.

Fiscal Deficit will increase as govt. have to do committed expenditure and revenue has fallen to almost zero because of prolonged lockdown.

The budgeted expenditures risen because of the steep rise in subsidies to workers and a sharp fall in all taxes, the Fiscal Deficit will rise dramatically, unless the planned expenditures are cut back.

Income, savings, investment, revenue of the government, tax GDP ratio, all macro-economic variable are falling constantly in Indian economy. There is need of a paradigm shift in the field of economics to tackle the post lockdown economic problems, as happened in economic history after the great depression of 1930.When the great depression came in 1930, classical tools failed to solve of economic problems of depression. Then J.M.Keynes observed that these micro tools cannot be applied on macro level problems, as a policy which is suitable for a person cannot be suitable for the whole economy so the desired results were not coming and after it Keynesian era started and there was a paradigm shift in economics. Now, the whole world is facing a worldwide depression because of spread of pandemic.We are standing on lowest point of trade cycle waiting for stimulants of revival but extent of income reducing shocks are more powerful than stimulants and India and whole world are searching for solutions to achieve previous growth rate.

Abhijeet Banerjee a senior economist suggested government to give cash doles to poor people so demand can be generated and it will boost investment in the economy and on the other side finance minister released four trenches of loan for different sector, but results are awaited as we cannot say that economy will come out from this covid pause by this loan mela as we found that MSME's repaid their debt from this loans and about 25000 crore rupees went back to RBI.We need the accurate policies that can bring paradigm shift in economics. Many senior economist criticized this lending policy of RBI and suggested spending policy to boost demand.

In coming times all economies will prefer to invest a big part of GDP on health and medical facilities.In labour abundance economy like India our focus should be on labour intensive technology to return to the previous path but in a competitive world we cannot ignore interdependence of international trade as manufacturing sector of India depends on imports of comparatively cheap economic substitutes for the production of final goods. So derived demand of cheap economic substitutes for final products is major import item of India' total imports. For example;Gems and jewelry industries of Jaipur, car and pharmaceautical industries depends on cheap intermediate imports. Instead of ANBA and self -reliant India, our strategy should be of trade promotion and interdependence of international trade for uninterrupted supply chain of intermediate goods to make final product by our industries. International theories of trade also prove that international trade is the engine of economic growth.

Conclusion

80 crore people are given food security for months and now a central minister says that it is not possible to pay MSP as farmers have overproduced so they are thrown amid free market forces as market price of food grains is less than MSP. Moreover, they give argument that If govt.purchase all production on MSP it would cost 17 lakh crore which is half amount of union budget. Someone is playing with data. Food production is not a waste which will be thrown or dumped. Government would compensate the farmer only the difference between the market price and MSP, not the whole MSP. Wrong interpretation of the data as according to the whole MSP of the cropcreates confusion among common men.Moreover, the farmer is producing to feed the population of nation. The question arises that if you have enough to eat than why GHI is going down and even ranked below Bangladesh and Pakistan. Indian people know hunger well. A food grain surplus nation cannot rank as low as India in the Global Hunger index. The common myth that we are the huge producers of foodgrains and export a big amount of foodgrains may be because we are exporting more and consuming less. The per-capita consumption of grains started declining in 1990 and this trend continued in later two decades and a recovery in 2011 -2020 and unfortunately this decline was not taken seriously as academicians interpret it according to Lorenge curve that with increasing income a consumer diversifies his consumption and spends a lesser amount on consumption of foodgrains and more on fruits, nuts and animal products. But in case of India this doesn't happened as proven by GHI and malnutrition among children.If demand disappeared or declined for foodgrains, where did the grain go ?. For the last

ten years India has emerged as a big cereal exporter and we are not food surplus state but consume less foodgrains. There is no authentic surplus food-grain in the country. Any surplus visible in the forms of large exports and overflowing FCI stocks can be attributed to the fact that the average Indian was not able to maintain her already low level of grain consumption over the last thirty years. All myths regarding 'food surpluses' are really based on the dystopia of the endless pangs of hunger, under-nutrition and their consequences that the average Indian and her children have borne over the last thirty years. Needless to say, this 'average Indian' is located far away from the glittering lives of the metropolitan cities and invisible from the world of policy-makers. 196 million people are undernourished and malnutrition is the top cause of deathand disability. It is clear that food security remains a serious challenge. India now ranks 94th out of 107th countries in terms of hunger and continue to be in 'severe' hunger category according to the GHI of 2020. These situations clearly proves that food is not accessible in India. South Asia has the highest child wasting rate for any region, and India is the worst performer. At 17.3%, India's child wasting rate is only slightly better than it was last year, at 20.8%.

Sometimes in present scenario I feel that the government is more worried about marketing of agricultural output instead of production. The focus should be on production and diversification of crops. Farmer face two types of risks,one is of marketing and second is of adverse weather. To cover the risk of marketing MSP should have a legal provision and for unfavorable weather condition the insurance of crops should be done by the government. Farmers should be encouraged to do crop diversification and to produce commercial crops. For diversification and commercialization more crops should come under MSP. To increase the income of this sectorArgo processing centers should be established at village level and Storage godowns at panchayat level and the farmers should be made stakeholders in these processing units. Agriculture should be promoted as an profession instead of an occupation and a farming should be made as an entrepreneurship.In China 1% increase in agriculture growth reduces 0.7% income inequality.So can be achieved in India if we will not ignore the backboneof economy.

Government has started many schemes to uplift this sector. Currently, there is a cash or income support schemes. The 2020-21 Union Budget allocated Rs 750 billion ($10.6 billion) for the direct income transfer scheme PM-KISAN.

However, the latest OECD assessment says that India is one of the few countries that has penalized farmers to keep consumers happy. The international measure of a government's budgetary and other subsidies to farmers is the Producer Support Estimate (PSE), developed by OECD that uses this for its annual tracking of global agriculture supports. In simple terms, this measure estimates what a farmer receives at the farm gate. OECD assesses that it is negative 5.7 per cent for Indian farmers, or that the government has actually taxed the farmers. In 2019, Indian farmers lost $23 billion this way. In contrast, Norway offers 60 per cent support to its farmers. India's negative PSE benefits the consumers in term of cheaper food or our obsession to control wage good inflation. Mostly, government support and policy intervention keep the wholesale price low and also help distribute cheap produce through the public distribution system to keep food inflation low. This way, consumers gained a benefit of $80 billion.

This is opposite to what most countries do: Keep agricultural produce price higher than the global level and make the consumers and government supports compensate for itas done in China. This means a good return to farmers. Even America highly subsidize it's farmers.

Government support to producers has remained negative in the last two decades in India. While allocations for fertilizer and food subsidies increased between 2018 and 2019, both fertilizers and food subsidies were lowered for financial year 2020-21 (by 10.8 per cent and 37 per cent, respectively).

Indian farmers suffered a cumulative loss of Rs 45 lakh crore ($ 600 billion plus) between 2000 and 2016-17 on account of being denied their rightful income. If we calculate backward working of income reducing effect on economy than this figure may be very high. No doubt the loss of farmers is someone's gain as of poor consumers$ 80 billion. No doubt the income of rich corporates has increased because of this loss of farmer and this income is going to be converted in black money and will increase income inequalities among sectors. Indian economy is going towards unbalanced growth by neglecting primary sector. According to new economic policy the target is to reduce the burden of population on primary sector but it has done no preparation for this as where the surplus working population will be adjusted and without a strong rural infrasture.

NITI Aayog itself had admitted that between 2011-12 and 2015- 16, the growth in real farm incomes had prevailed at less than half a per cent every year, 0.44 per cent to be exact. For the next two years, the growth in real farm income had been 'near zero', according to agriculture expert Devender Sharma.

Conclusion

The small and marginal farmers account for 85 per cent of the total landholdings and hold close to 40 per cent share in the total 'marketable surplus.

The small and marginal farmers will be ruined by the market interventions. Markets are complex institutions representing economic relationships embedded in the prevailing socio-political realities. The price formation in a particular market is dependent on the demand and supply dynamics as well as a myriad of structural factors. The balance of bargaining power between any two parties determines the price formation in a commodity market. Consequently, a floating demand does not necessarily guarantee a better price for the seller if the buyer is a monopolist. Similarly, the absence of a monopoly buyer does not guarantee high prices to the seller if the demand is slow or declining. The demand and supply itself is influenced by a number of factors that are often outside the realm of market transactions.

With globalization and a revolution in transportation, the market demand and supply are influenced by so many number of factors. For agricultural commodities, natural and weather-related events further add to the complications of the demand-supply dynamics.

This has meant far greater fluctuation in agricultural commodity prices after trade liberalization. There have been periods in history when prices have boomed but frequently followed by huge slumps wiping out past gains for small to big farmers. In light of this, any blind claim that deregulating markets and allowing the free play of private players will improve crop prices for farmers appears to be on unstable grounds at best, and spurious at worst.

The most important argument that is not considered when it is argued that removing the arhatiyas (middlemen) will benefit the farmers is the structure of global value chains in agricultural commodities. It is widely accepted now that agricultural value chains typically take the shape of an hour-glass, with large number of producers and consumers having negligible bargaining power at the two ends and a few wholesalers, processors and retailers which control all the levers in the commodity markets. The arhatiyas and the government supervision/regulation provide a reasonable bargaining power and call for accountability to the farmers, which global value- chains have systematically denied to farmers across the world. Moreover the commission paid to APMC is spent on rural infrasture by the state government. That is the reason of a solid rural infrasture in rural Haryana and Punjab. There is no need of double mandi system as P.M. said one private and other APMC's. This will dangerous to throw the farmers among private players and corporates and it will have spillover effects on society in the long run. There is every possibility that the private buyer will export the wheat on comparatively high prices and there will be shortage of food grains in domestic market.

Modi government brought these bills at the peak time of CORONA pandemic. Not enough time was given to farmers and this created confusion, fear and distrust among farmers toward government. In economics we read that the producer react to price and plan production accordingly and assure price is best incentive for farmers.To keep the demand and supply matching State government can collect demand estimates of food grains and can ask to farmers plan production according to demand so that there will be no excess supply of foodgrains and than market price will not fall below a certain level. Farmer will be assure about the future demand and will produce accordingly by adopting a diversification method.Food grain should be sufficient and accessible for society.

In India the situation of agriculture and farmers are so degradable and pathetic that if farmers will have other option for their livelihood than about 45% farmers will leave this occupation. As a remedy for this and to save the lifeline of the economy the budget expenditure on rural infrasture should be increased and storage godwon and Argo processing units can be started at village level and labour movement can be stopped than no one will leave its birth place if they will earn enough to feed their family. We have to generate work for a majority of people as an additional source of income by setting up Agro processing units and a huge investment in rural infrastructure so that each villager will get employment. Same can be done by making farmers stakeholders in agro processing units and the producer will get share in value addition by these processing units. Loan by banks and should be kept secret instead of declaring them defaulter they should be given time to repay in easy installments.

About 195 million hectares is total arable land in India and from this only 38% have irrigation facilities. Efforts should be done to increase the irrigated land. According to the data of NCRB, lack of irrigation facilities is one of the major reasons of increasing suicide rate among farmers of Maharashtra. According to Yogender Yadav the head of Swaraj India that there are three main intertwined crisis in agriculture of India. Economical, ecological and a newly created crisis by central government of India, the crisis of existence. We have already discussed economic crisis that the income of a farmer is so meagre that it has become difficult for him to lead a respectful life, hardly 7000 to 8000 per month. Ecological crisis came after green revolution in the form of degradation of agriculture land because of the use of chemical fertilizers and depleting water level. Crisis of existence of farming community is coming crisis when the farmer will loose his arable land because of neglecting policies of government. Farmers have fear that they will also going to swell the army of landless labour in near future. To overcome these crisis in India we have to design a new growth model for this sectors with a assured price of wheat and paddy, many more crops should be added under MSP and then the farmer will get an incentive to produce according to the ecological needs of the people and in this model there will be no overproduction of one or two crops and there will be no need of procurement of crops as production will be according to demand. Side by side rural industries should be established by the government by making farmers stakeholder. Our handicraft and jute products have a large demand in international markets, so these lost industries should be given a boost and permotion. India is a country of various geographical conditions and our products have demand in international market as basmati rice, turmeric, various spices and gur made from sugarcane. India can export these unique products in international market. Farmers should be financed by banks for this purpose and government should decide portfolio of loans to farmers in a transparent way for export promotion and for this change in crop diversification with a guarantee of MSP, so that the agriculture can be made a profession of profits and farmer an entrepreneur.

Author's Bio

Dr. Harvinder Kaur is presently working as Assistant Professor in Economics in S.D.College, Ambala (Haryana). She was conferred Ph.D Degree in 2004 in Agriculture Economics from Kurukshetra University, Kurukshetra. She has been awarded National scholarship in school, graduation and research fellowship for her Ph.D from K.U.K. Her specialization is in Econometrics and research interest is in Agriculture Economics and Macro Economics. She holds more than thirteen years of teaching experience and has attended more than forty national and international seminars and got published about twenty papers in national, international journals, newspapers and has edited a book on Food Security. She has also organized seminars in the field of economics. She is member of Haryana Economic Association and also have a strong passion of social work, as she is member of two Kurukshetra based NGO's which work for welfare of girl child. She has a spiritual bent of mind and organize meditational camps time to time.

E-mail: hksanjivani777@gmail.com

www.ingramcontent.com/pod-product-compliance
Lightning Source LLC
Chambersburg PA
CBHW020926180526
45163CB00007B/2897